PUBLICATIONS OF THE
NATIONAL BUREAU OF ECONOMIC RESEARCH, INC.

NUMBER 32

NATIONAL INCOME
AND CAPITAL FORMATION
1919–1935

Officers
Joseph H. Willits, Chairman
George Soule, President
David Friday, Vice-President
Shepard Morgan, Treasurer
Joseph H. Willits, Executive Director
Martha Anderson, Editor

Directors at Large
Chester I. Barnard, *President,
New Jersey Bell Telephone Company*
Henry S. Dennison, *Dennison Manufacturing Company*
George M. Harrison, *President,
Brotherhood of Railway and Steamship Clerks*
Oswald W. Knauth, *President, Associated Dry Goods Corporation*
Harry W. Laidler, *Executive Director,
The League for Industrial Democracy*
L. C. Marshall, *Johns Hopkins University*
George O. May, *Price, Waterhouse and Company*
Shepard Morgan, *Vice-President, Chase National Bank*
Beardsley Ruml, *Treasurer, R. H. Macy and Company*
George Soule, *Director, The Labor Bureau, Inc.*
N. I. Stone, *Industrial and Financial Consultant*

Directors by University Appointment
William L. Crum, *Harvard*   Harry Alvin Millis, *Chicago*
Walton H. Hamilton, *Yale*   Wesley C. Mitchell, *Columbia*
Harry Jerome, *Wisconsin*   Joseph H. Willits, *Pennsylvania*

Directors Appointed by Other Organizations
Frederick M. Feiker, *American Engineering Council*
David Friday, *American Economic Association*
Lee Galloway, *American Management Association*
Malcolm Muir, *National Publishers Association*
Winfield W. Riefler, *American Statistical Association*
George E. Roberts, *American Bankers Association*
Matthew Woll, *American Federation of Labor*

Research Staff
Wesley C. Mitchell, Director
Arthur F. Burns   Frederick R. Macaulay
Solomon Fabricant   Frederick C. Mills
Simon Kuznets   Leo Wolman
Eugen Altschul, David L. Wickens, Associates

RELATION OF THE DIRECTORS TO THE WORK OF THE NATIONAL BUREAU OF ECONOMIC RESEARCH

1. The object of the National Bureau of Economic Research is to ascertain and to present to the public important economic facts and their interpretation in a scientific and impartial manner. The Board of Directors is charged with the responsibility of ensuring that the work of the Bureau is carried on in strict conformity with this object.

2. To this end the Board of Directors shall appoint one or more Directors of Research.

3. The Director or Directors of Research shall submit to the members of the Board, or to its Executive Committee, for their formal adoption, all specific proposals concerning researches to be instituted.

4. No study shall be published until the Director or Directors of Research shall have submitted to the Board a summary report drawing attention to the character of the data and their utilization in the study, the nature and treatment of the problems involved, the main conclusions and such other information as in their opinion will serve to determine the suitability of the study for publication in accordance with the principles of the Bureau.

5. A copy of any manuscript proposed for publication shall also be submitted to each member of the Board. If publication is approved each member is entitled to have published also a memorandum of any dissent or reservation he may express, together with a brief statement of his reasons. The publication of a volume does not, however, imply that each member of the Board of Directors has read the manuscript and passed upon its validity in every detail.

6. The results of an inquiry shall not be published except with the approval of at least a majority of the entire Board and a two-thirds majority of all those members of the Board who shall have voted on the proposal within the time fixed for the receipt of votes on the publication proposed. The limit shall be forty-five days from the date of the submission of the synopsis and manuscript of the proposed publication unless the Board extends the limit; upon the request of any member the limit may be extended for not more than thirty days.

7. A copy of this resolution shall, unless otherwise determined by the Board, be printed in each copy of every Bureau publication.

(Resolution of October 25, 1926, revised February 6, 1933)

SIMON KUZNETS

NATIONAL INCOME AND CAPITAL FORMATION 1919-1935

A PRELIMINARY REPORT

NATIONAL BUREAU OF
ECONOMIC RESEARCH
NEW YORK · 1937

COPYRIGHT, 1937, BY

NATIONAL BUREAU OF ECONOMIC RESEARCH, INC.

1819 BROADWAY, NEW YORK

ALL RIGHTS RESERVED

DESIGNER: ERNST REICHL

MANUFACTURED IN THE U. S. A. BY

H. WOLFF, NEW YORK

FOREWORD

THE measures of national income and capital formation presented in this volume resulted from three studies that have been carried on at the National Bureau almost independently: national income and capital formation under my direction—the latter undertaken at the request of the Committee on Banking and Credit of the Social Science Research Council; and capital consumption under Solomon Fabricant's direction. Since the magnitudes that these three studies measure are closely related it seems advisable to present the preliminary summary of the first two, national income and capital formation, in a single volume, and make complementary use of Mr. Fabricant's estimates of capital consumption.

The measures of national income constitute a complete revision of the estimates for 1919–28 by W. I. King (*National Income and its Purchasing Power*), and their continuation through 1935. In utilizing the estimates prepared by the Department of Commerce in cooperation with the National Bureau for 1929–32, as revised and brought up to date by the former agency, we were permitted access to unpublished details.

The measures of commodity flow and capital formation constitute a revision and amplification of preliminary estimates of somewhat narrower scope released in Gross Capital Formation, 1919–1933 (*Bulletin 52*, National Bureau of Economic Research, November 15, 1934).

Some of the national income measures have been published for circulation among National Bureau subscribers in National Income, 1919–1935 (*Bulletin 66*, National Bureau of Economic Research, September 27, 1937). The estimates presented below incorporate minor revisions made since that was written. These revisions are due to slight changes in the measure of capital consumption by government and in the data on building permits just released by the United States Bureau of Labor Statistics.

Detailed description of the sources and methods used in obtaining the estimates of national income and capital formation will appear in the final report on national income, and in Volume I of the final report on *Commodity Flow and Capital Formation*, which is now in press. These final reports may well include further improvements suggested by later data. But such changes are not likely to affect sensibly the estimates in this volume.

In the preparation of the estimates of national income the writer has been assisted by Lillian Epstein and Elizabeth Jenks from the beginning of the study, and by Edna Ehrenberg during recent months. Throughout the investigation of capital formation assistance was rendered by William H. Shaw; and for various intervals by Grace W. Knott, Edith Handler, Lawrence Robinson, Richard Machol and Natalie Brown. The writer is indebted for help also to the research staff of the National Bureau, particularly to Solomon Fabricant who prepared the estimates of capital consumption, and to Arthur F. Burns whose suggestions led to a revision of the original manuscript.

SIMON KUZNETS

November 3, 1937

CONTENTS

Foreword	vii
Introduction	1

I DEFINITIONS AND SCOPE — 3
1. Gross and Net National Product — 3
2. The National Product at Successive Stages of Economic Circulation — 4
3. Scope of the Estimates — 5
4. Limitations of the Estimates — 6

II TOTAL NATIONAL PRODUCT — 8
1. The Totals and Their Changes — 8
2. Comparison with Population Changes — 10

III DISTRIBUTION ACCORDING TO INDUSTRIAL ORIGIN — 12
1. Meaning of Classification — 12
2. Distribution in Current Prices — 13
3. Distribution in 1929 Prices — 20

IV DISTRIBUTION ACCORDING TO TYPE OF INCOME — 23
1. Meaning of Classification — 23
2. Importance of Savings of Enterprises — 26
3. Changes in Distribution by Type of Payment — 27

V CROSS-CLASSIFICATION BY TYPE OF PAYMENT AND BY INDUSTRIAL SOURCE — 29

VI THE DISTINCTION BETWEEN CONSUMERS' OUTLAY AND CAPITAL FORMATION — 34
1. National Product, Consumers' Outlay and Capital Formation — 34
2. Methods of Estimation and the Classification of Commodities — 35
3. Variants of the Distinction between Consumers' Outlay and Capital Formation — 37

VII APPORTIONMENT OF GROSS NATIONAL PRODUCT BETWEEN GROSS CAPITAL FORMATION AND CONSUMERS' OUTLAY — 39
1. Gross Capital Formation—Characteristics of the Estimates — 39
2. Volume and Composition of Gross Capital Formation — 41
3. Apportionment between Gross Capital Formation and Consumers' Outlay — 44

VIII APPORTIONMENT OF NATIONAL INCOME BETWEEN NET CAPITAL FORMATION AND CONSUMERS' OUTLAY — 47
1. Net Capital Formation, Volume and Composition — 47
   a. Characteristics of the estimates — 47
   b. Comparison with gross capital formation and capital consumption — 49
   c. Absolute volume compared with wealth — 50
   d. Distribution among component elements — 51
2. Apportionment of National Income between Net Capital Formation and Consumers' Outlay — 52

IX THE COMPOSITION OF CONSUMERS' OUTLAY — 55

X SUMMARY — 58

APPENDIX
A. National Income and Aggregate Income Payments to Individuals by Type of Income and Industrial Branch, Basic and Other Variants — 61
B. Comparison with Department of Commerce Estimates — 76
C. Changes in Procedure and Scope since the Publication of Preliminary Estimates of Capital Formation in Bulletin 52 — 78
   1. Flow of Finished Commodities — 78
   2. Volume of Construction — 79
   3. Repairs, Maintenance and Servicing — 81
   4. Net Changes in Commodity Inventories — 81
D. Comparability of Estimates of Capital Formation with Those of National Product — 82

TABLES

1 National Income and Gross National Product, 1919–1935   8
2 Distribution of Gross National Product According to Industrial Origin, 1919–1934   14
3 Distribution of National Income According to Industrial Origin, 1919–1935   16
4 Distribution of Aggregate Income Payments to Individuals According to Industrial Origin, 1919–1935   18
5 Distribution of Gross National Product and National Income, in 1929 and Current Prices, by Origin in Commodity Producing and in Other Industries, 1919–1934   21
6 Distribution of Gross National Product and National Income According to Type of Income, 1919–1935   24
7 Change in Percentage Share of Various Types of Payment within Each Industrial Branch   30
8 Change in Percentage Share of Various Types of Payment in Aggregate Income Payments to Individuals, Total and Resulting from Intra-Industry Shifts in Types of Payment and Inter-Industry Shifts in Total Payments   32
9 Employees' Compensation and Aggregate Income Payments, Percentage Distribution among Industrial Branches   33
10 Gross Capital Formation, by Type of User, 1919–1935   40
11 Average Volume and Distribution of Gross Capital Formation   42
12 Apportionment of Gross National Product between Gross Capital Formation and Consumers' Outlay, 1920–1934   45
13 Net Capital Formation, 1919–1935   48
14 Average Volume of Gross Capital Formation, Capital Consumption and Net Capital Formation   50
15 Apportionment of National Income between Net Capital Formation and Consumers' Outlay, 1920–1934   53
16 Composition of Consumers' Outlay, 1920–1934   56
17 Average Value of Significant Items   58

APPENDIX TABLES

I National Income and Aggregate Income Payments to Individuals by Industrial Branches, Basic Variant, 1919–1934   62
II National Income and Aggregate Income Payments to Individuals, Other Variants, 1919–1934   68
III Estimated Losses and Gains by Business Enterprises on the Sale of Capital Assets, 1929–1934   75
IV Comparison of Aggregate Income Payments (N.B.E.R.) with Income Paid Out (D. of C.), 1929–1935   77
V Comparison of Net Saving of Enterprises (N.B.E.R.) with Business Savings (D. of C.), 1929–1935   77
VI Comparison of Total National Income (N.B.E.R.) with National Income Produced (D. of C.), 1929–1935   78
VII Value of Repairs, Servicing and Maintenance of Durable Commodities, 1919–1933   80
VIII Annual Estimates of National Product, Capital Formation and Consumers' Outlay, 1919–1935   85

CHARTS

1 National Product, 1919–1935   9
2 Percentage Distribution of Gross National Product, National Income, and Aggregate Income Payments by Major Industrial Divisions, 1919–1935   20
3 Percentage Apportionment of Gross National Product, and National Income, in 1929 Prices, by Origin in Commodity Producing and in Other Industries, 1919–1934   22
4 Various Types of Payment and Savings of Enterprises in Percentages of Aggregate Income Payments, 1919–1935   28
5 Percentage Distribution of Aggregate Income Payments among Major Types of Payment, 1919–1934
Total and Resulting from Intra- and Inter-Industry Shifts   31
6 Gross Capital Formation, 1919–1935   43
7 Percentage Apportionment of Gross National Product between Gross Capital Formation and Consumers' Outlay, 1920–1934   46
8 Net Capital Formation, 1919–1935   49
9 Percentage Apportionment of National Income between Net Capital Formation and Consumers' Outlay, 1920–1934   54
10 Composition of Consumers' Outlay, 1920–1934   57

NATIONAL INCOME
AND CAPITAL FORMATION
1919–1935

INTRODUCTION

This report seeks to answer five questions: (1) How can the product of the nation's economic activity be defined? (2) How large has the national product been since the War? (3) In what part of the nation's economic system was it produced? (4) How was its monetary equivalent distributed? (5) How was the product utilized?

Question one is answered briefly in Section I, which discusses definitions and scope. In Section II estimates of the total national product are presented and changes in it compared with changes in population. The third question is answered, in Section III, by a study of the distribution of the national product according to industrial origin. The distribution of the monetary equivalent of the national product by type of income is studied in Sections IV and V. The measurable distribution by type of utilization is first discussed in Section VI, in which the component elements of consumers' outlay and of capital formation are indicated. The apportionment of the national product between capital formation and consumers' outlay and the component elements within each are then studied in Sections VII through IX.

The measures of national income, of the related total—gross national product—and of the various distributions of these totals, lend themselves to extensive use in economic analysis. The estimates of the total national product, if properly interpreted, indicate the volume of goods yielded by economic activity and available either for immediate consumption or for additions to the stock of capital. Such measures of the current supply of goods provide totals with which one can compare various possible drafts, significantly different uses to which these goods may be put, or significantly different ways in which they have been or could have been produced. The distribution according to industrial origin is useful in suggesting changes in the industrial structure of the economy. The distribution by type of income is useful in indicating, first, how much of the total net value produced is retained by business enterprises and how much is distributed to individual income recipients, thus becoming a source of their expenditures and savings; second, how much of the latter total, aggregate income payments to individuals, is paid in various forms (wages, salaries, dividends, interest) that indicate tentatively, if not definitively, the distribution of payments among income classes distinguished according to the size of their average income. The apportionment between consumers' outlay and capital formation is of obvious use in economic analysis, in view of the significant difference in the behavior of those two closely related segments of the economic system. Similarly, the distinction among the various components that make up capital formation and consumers' outlay can be used to reveal those groups in the total product whose changes over time are likely to be significantly different, and whose movements are the result of different groups of forces that must be distinguished if economic processes are to be understood.

It is as approximate indicators of the magnitudes of the significant areas in the economic system, of the relation of these magnitudes, and of the broad changes in these relations from 1919 through 1935 that the measures are primarily useful. They are subject to limitations that bar their utilization for types of analysis that demand precise data, recording changes over short periods. For example, they can be but rough guides in the analysis of business cycles, owing to the tentative character of some of the estimates that make up the national product totals, as well as their annual character. Also, a complete understanding of even the broader movements is sometimes rendered difficult by the shortness of the period studied, 1919–35; and even for this period the lack of available data makes it impossible to check certain observations by a more intensive analysis.

The possible uses and the present limitations of the measures will be revealed more clearly in the discussion, which attempts to bring out their significance as a summary statement of the end-product of the functioning economic system. We have not attempted to combine these measures, which are the result of several studies carried on at the National Bureau of Economic Research, with other data available elsewhere, in order to discuss even a

INTRODUCTION

few of the vast range of economic problems to whose solution they might contribute. In this sense, this report is merely a preliminary summary of the results of the studies of national income, of capital formation, and in less degree, of capital consumption. But the results, even in themselves, provide a useful picture of the economic scene in this country since the World War. And so far as the scope and limitations of the measures are made clear in subsequent discussion, the way is open for their use, in combination with other data, for the analysis of specific economic problems.

I DEFINITIONS AND SCOPE

1 GROSS AND NET NATIONAL PRODUCT

NET national product or national income may be defined as the net value of commodities and services produced by the nation's economic system. It is 'net' in that the value of output of all commodities and services is reduced by the value of commodities (fuel, raw materials and capital equipment) consumed in the process of production. And it refers, by design, to the net product of the economic system, which, for economically advanced nations of recent times, may be treated as identical with the market economy. If 'market' is understood broadly as the meeting place of all buyers and sellers, no matter how greatly freedom of transactions may be curbed by custom or by regulation, then national income measures the net product of activities connected through the market, and excludes the results of other activities that may supply utilities but are outside the market mechanism.

Interpreted thus as the net current supply of goods available either for immediate consumption or for addition to the already existing stock of wealth, national income is a uniquely determined concept. But other comprehensive measures may be conceived wherein the product is 'gross' in that it is not fully adjusted for the value of commodities consumed in the process of production. For example, it is possible and sometimes proper, in measuring the result of economic activity, to count the value of both the coal and the machine in whose production the coal was consumed; to include both the automobile and the machine tools used in making it, although the latter in each case were at least partly consumed in turning out the former. Thus, in contrast to the single concept of net national product, there may be several concepts of gross national product, which vary with the amount of duplication involved. Such duplication may be allowed among separate plants, separate enterprises, separate industries, or major industrial branches of the economic system—yielding several gross totals of the national product, each substantially different from the others. The magnitude of these gross totals will vary with the amount of duplication desired and obtainable with the available data.

Of these several possible concepts of gross national product one appears of greater importance than the others, that in which the value of commodities and services produced is not adjusted for the value of durable capital goods consumed in the process of production, but is adjusted for raw materials, partly fabricated products and fuel consumed. It is this concept that is referred to below as gross national product and is measured in the tables. The reason for singling out this particular concept of gross national product lies in the peculiar nature of the durable [1] commodities that constitute capital equipment. It is characteristic of durable commodities that they are modified only gradually during the process of consumption. In contrast to the consumption of raw materials, which is a clearly apprehended and measurable process, the current consumption of durable commodities is at best a rough estimate, even when its evaluation is attempted by the productive or consumptive agency that uses the durable goods in question. Hence gross national product has the advantage of being a variable that can be measured more accurately than net national product. More important is the fact that the replacement of durable capital goods in use by new commodities is not as rigidly controlled by technical considerations as is the replacement of raw materials; considerable discretion is left to business enterprise. This means that over short periods the stock of capital equipment may be treated as indestructible, and its consumption in the process of production neglected. On this assumption the supply of commodities and services resulting from economic activity and at the disposal of the nation during any brief period includes not merely the net value of commodities and services produced, but this net value augmented by the deductions made for the presumptive current consumption of durable capital goods. Consequently, in addition to net national product or national income, we also measure gross national product, i.e., the volume of commodities and services produced, corrected for all duplications in fuel, raw materials, semifinished and finished products, but not reduced by deductions for the current consumption of capital equipment.

[1] For purposes of this study durable commodities are those whose period of utilization is more than three years; see Section VI.

2 THE NATIONAL PRODUCT AT SUCCESSIVE STAGES OF ECONOMIC CIRCULATION

The above definitions describe the net and gross results of economic activity at the point of production, referring as they do to the values of commodities and services *produced*. But in the circuit flow of goods and money that constitutes the functioning of our economic system, the yardstick that would provide a comprehensive measure of the results of economic activity may be used at any stage. National income and gross national product are produced, distributed and consumed, and it is possible to define and measure them not only at the point of production but also at the point of distribution and of consumption.

At the point of production, national income, so far as it represents the net result of economic activity, represents that part of gross national product which is largely imputable to individuals who contribute to production either their labor or the services of their property. The compensation received by these individuals in return for this participation accounts for the preponderant share of national income produced in any given year. But there may be two other elements in national income. First, the producing enterprises may make payments not only to individuals who, in a given year, participate in the productive process, but also to individuals who participated in it in the past or have not participated at all. Pensions, compensation for injuries, relief payments, and charity contributions by business enterprises are channels through which income produced in a given year may be paid to individuals other than those sharing in the process of producing it. Second, enterprises may distribute to individuals amounts not necessarily equal to the part of the national income that these enterprises produce. In some years a business firm or other producing enterprise may pay out to individuals an amount smaller than its share in national income, thus retaining what we designate as positive net savings of enterprises. In other years an enterprise may distribute to individuals an amount larger than its share in the national total, thus giving rise to negative net savings of enterprises. If we designate all receipts by individuals from the producing enterprises as income payments to individuals, national income is equal to the algebraic sum of income payments to individuals and of net savings of all enterprises.[2]

Gross national product includes, besides income payments to individuals and net savings of enterprises, the amounts deducted by the users of durable capital goods as an allowance for the current consumption of these goods in the productive process. If these deductions, which appear largely as depreciation and depletion charges, are added to net savings of enterprises, the resulting total represents the share of gross national product retained by enterprises. If we designate this share as gross savings of enterprises, then gross national product, at the point of distribution, is equal to income payments to individuals plus gross savings of enterprises.

Furthermore, the individuals who comprise a nation are ultimate consumers, and the value of the products they consume during a given year may be larger or smaller than the aggregate of incomes paid them. The difference between aggregate incomes paid to individuals and the value of products consumed by them represents individuals' savings, which may be either positive or negative. Consequently national income may be defined, third, as an algebraic sum of the value of the products that individuals consume, plus their aggregate savings, plus net savings of enterprises. Likewise, gross national product is equal to the value of the products consumed by individuals, plus their aggregate savings, plus gross savings of enterprises.

The three definitions indicate that national income and gross national product may be represented as a cross-section at any stage in the circulation of economic goods—production, distribution or consumption—with results that, if no statistical difficulties are met, should be identical.[3] But equa-

---

[2] National income as defined here is identical with what in earlier publications we designated as 'national income produced' (see *National Income, 1929–1932*, Senate Doc. 124, 73d Cong., 2d Sess., and *Bulletins* 49 and 52 of the National Bureau of Economic Research). The magnitude designated in these publications as 'national income paid out' is here identified as 'aggregate income payments to individuals'. This change in terminology resulted from the discussions prior to and at the meeting of the Conference on Research in National Income and Wealth held in January 1937. It was concluded that the use of the two terms 'national income produced' and 'national income paid out' as equally important concepts created confusion and led to misinterpretations; that the term 'national income' should be reserved for the magnitude formerly called 'national income produced', as being the most inclusive category and most consonant with the national income concept in economic writings; and that the magnitude formerly called 'national income paid out', being a subdivision of national income, would be best described by a specific term. For a more extended discussion see Part One of *Studies in Income and Wealth*, Volume I, by the Conference on Research in National Income and Wealth (National Bureau of Economic Research, 1937).

[3] This identity is true for any time unit for which the national product is measured, provided that we conceive national product properly as a sum of values rather than identify it improperly with a congeries of specific commodities and services. The

tion of the national product, whether net or gross, as determined by the three definitions, is valid only if its various elements are defined consistently. If we exclude the results of certain activities (for example, gambling) from the first definition, we cannot, when we use the second definition, include in individuals' incomes the compensation received from these activities, nor, when we use the third definition, can we include in the value of products consumed the expenditures on these activities. Such judgments as to what constitutes commodities and services, income shares, consumption, savings, etc., are of telling importance in the measurement of the national product. They are based, consciously or unconsciously, upon concepts of economic activity and productivity implicit in all measures. We cannot discuss here the numerous decisions based on these judgments, but it is essential to indicate how, in making the most important, the scope of the measures presented below was determined.

3 SCOPE OF THE ESTIMATES

The available data render it necessary to estimate national income by a procedure corresponding to the second definition, i.e., the aggregate of all incomes paid to individuals plus net savings of all enterprises. Likewise, gross national product is estimated by adding to national income the amounts previously deducted as representing the current consumption of durable capital goods. We therefore indicate the scope of the measures in terms of the second set of definitions. Obviously the limits established by it automatically indicate the scope of the national product in terms of either of the other two sets of definitions.

As already stated, national income and gross national product measure the result of activities that are closely connected with the market system. With respect to that part of the national product which was designated as income payments to individuals, the natural procedure would be to include only those shares that are received on the market in the form of money. This would mean including such payments as wages, salaries, dividends and interest, and excluding returns to individuals from their activity within the family system (housewives' services, etc.), or from other pursuits that are not, strictly speaking, working for the market. While such limitation is, on the whole, acceptable, there are a few types of income that do not pass through the market and do not find their equivalent in the sphere of monetary circulation, but that should nevertheless be included as representing the result of economic activity: (1) the share of the product retained by entrepreneurs for their own consumption (largely on farms); (2) payments in kind to employees (a practice especially important in compensating farm laborers, domestic servants, steamer crews, etc.); (3) services of houses inhabited by their owners. Except for these three groups of non-monetary payments, covered as adequately as possible with the data available, national income and gross national product include only such individuals' shares as are paid in money and constitute a part of the monetary circulation in the economy.

However, not all compensation to individuals for marketable activities is included, for a considerable number of such activities cannot be adjudged productive by the most lenient standard of productivity. For example, 'racketeering' services are obviously rendered in return for compensation, at prices determined by the balance of supply and demand on the market for such services. They are thus part of our economic system; but to include them in the national product totals would make it impossible to interpret the latter as the contribution of the economic system that appears useful to society at large. Such a criterion of productivity is unavoidable if estimates of the national product are not to become purely superficial measures of one aspect of the market mechanism.

In general, the national product totals presented below exclude the returns from three sources: (1) activities that have been explicitly recognized by society at large, overtly in the form of legal prohibition, not only as unproductive but also as distinctly harmful: theft, robbery, prostitution, murder, smuggling, counterfeiting, forgery, drug peddling; (2) activities that while legal, represent largely shifts of income among individuals rather than additions to the command over goods: gambling profits and speculative gains from the sale of assets by non-professional groups that can in no way be interpreted as resulting from skill in the performance of any useful functions that speculative

---

income payments resulting from the specific commodities and services produced in one time unit may be disbursed in another time unit, and the commodities and services themselves consumed in a still different time unit. But when the national product is conceived as a sum of values, the total value of goods produced *must* be equal to income payments to individuals plus savings of enterprises, and to consumption by individuals plus savings of individuals and of enterprises; since whatever values have been produced must either be paid out to individuals or retained by enterprises, and whatever income payments were received by individuals must either be spent on goods consumed or retained as individuals' savings.

markets may be presumed to render; (3) gains and losses that result from a mere change in the price level, on the same grounds that apply to the exclusion of speculative gains resulting from non-professional activity. However, while it was possible to exclude the direct results of these three sources of income, it was impracticable to eliminate their indirect effects on the valuation of income payments that were included as representing compensation for what is largely a productive activity.

The treatment of net and gross savings of enterprises should, of course, be strictly parallel to that indicated for income payments to individuals. Hence, with one exception, we include net and gross savings of only such enterprises as are engaged in buying and selling goods, savings that thus represent shares in commodities and services that appear on the markets of the nation. The single exception is the item of capital consumption charges on residential real estate inhabited by its owner. This item was included in gross savings of enterprises and gross national product although it is not a share of any current service appearing on the market. On the other hand, we excluded net and gross savings of enterprises engaged in illegal activities and attempted to exclude that element in the savings item which is due to changes in the price level.[4]

Finally, whatever results of economic activities were included in the national product totals were assigned the value they brought on the market. Thus the activity of a miner was valued at the wage he received. The services rendered by a capital investment were measured by the interest or dividends that the investor received. Clearly, certain peculiarities of the market place assign the various activities values other than those they would have within the framework of a different social system, or as measured from the standpoint of a social philosophy different from that of the market place. But the investigator who attempts to measure the national product in terms acceptable to society at large must use a valuation that, by its wide acceptance, represents approximately the valuation put by society on the various activities. So long as in our economic system the market place is the only mechanism by which the contributions of various individuals in various types of service are made commensurate, an estimate that is intended to be used widely must be referred to it. This does not affect the advisability of introducing corrections for some of the obscuring influences of market valuation upon the measurement of the command

[4] See discussion in Section IV.

over goods that national product totals represent. Of such adjustments the only one found practicable concerned changes over time in the general price level. In all other respects the various activities were valued at market prices, in spite of the differences in the valuation of the same type of activity from one market to another.

4 LIMITATIONS OF THE ESTIMATES

Estimates of the national product constitute a comprehensive gauge of the result of economic activity in our society. They are therefore useful in evaluating the contribution of the economic system to the needs of consumption and of capital accumulation, and constitute a comprehensive total, a general frame of reference, for studying changes over time in the productivity of the economic system and in the differences in the apportionment of the total product among various significant groups. These possible uses of the national product estimates will appear more clearly when we present them and attempt to indicate their significance. Here we must note the considerable limitations to which they are subject, and which must be taken into account in order to avoid common misinterpretations.

First, even if we disregard limitations imposed by lack of data, national income and gross national product do not measure all the goods and services produced in the nation, since they exclude, by design, a large volume of services and a substantial volume of commodities produced outside the economic system proper. The great contribution to our stock of utilities made within the family system and by numerous activities of mankind engaged in the ordinary processes of life is not included. Since the line between economic activity proper and other utility producing activities shifts from time to time and from country to country, this limitation of national product estimates to the result of economic activity proper should be kept in mind in any attempt to interpret them as approximations to the value of the total stock of commodities and services produced.

Second, even within these limits estimates of the national product do not necessarily measure adequately the result of economic activity in terms of command over commodities and services. As has been mentioned, the results of economic activity are evaluated at current market prices. These evaluations reflect, among other factors, inequalities in the distribution of income, differences among various types of service with respect to the competitive position of their producers, changes in the effective supply of money, and differences over space or time in the method of estimating the consumption of durable capital goods.

Third, the national product, even when we consider the part designated as aggregate income payments to individuals, does not represent all the means of purchase that are available to individuals during any given year. The power of individuals to buy in markets depends not only upon the current flow of income but also upon the possibility of exploiting other sources, such as assets and credit facilities. This suggests that measures of the national product, in order to be adequate as a gauge of the performance of the economic system even in terms of the market place, must be supplemented by a study of wealth and capital structure. Strictly defined, national product measures reflect only such changes in wealth as result from the disposition of the current flow of goods and services produced. Other important changes in the capital and wealth structure of a nation's economic system must be taken into account, in addition to those revealed by measures of the national product, in order to obtain a complete picture of the basic changes in the economic scene.

Besides these limitations there are the usual limitations arising from lack of reliable data for the accurate measurement of some constituent parts of the national product totals. These limitations could be understood clearly only from a detailed discussion of the sources and methods used in deriving the various estimates of which each national product total is an amalgam. Such a discussion cannot be entered upon in this report. In general, it may be said that lack of data limits the coverage of the product retained by entrepreneurs for their own consumption to farm produce retained by farmers for family consumption; prevents an adequate measure of all payments in kind to employees; leads to the exclusion of the small amount of imputed rent on owned farm homes; and results in an omission of a large number of incomes from odd jobs that are not recorded in the basic surveys of production and employment. Of the areas actually covered, the estimates are most reliable for the industrial branches for which basic data have been available for some time (mining, manufacturing, public utilities reporting to the Interstate Commerce Commission), and least reliable for the industrial branches for which basic Census data are still lacking or are of only recent origin (service industries, real estate, and, of course, the miscellaneous group). Of the various types of income the estimates are most reliable for those whose flow is recorded directly in the data (wages, salaries, dividends and interest), and least reliable for entrepreneurial withdrawals, rents and net savings of enterprises.

The effect of some of these limitations is reduced by the presentation of not only the national product totals, but also as many significant distributions as can be derived from the available data. Such allocations make it possible to distinguish the various areas of the economic system in which the peculiarities of the approach and the various inclusions and exclusions affect the estimates in different ways; and to indicate what particular parts of the estimates are especially tentative because of deficiencies in the data. But most important, the apportionments of the national product totals enable us to understand more clearly the character, causes and effects of the observed changes in these totals. For this reason, the discussion following Section II is devoted largely to a treatment of the various significant allocations made possible by the available data. After showing how national income and gross national product are distributed among the various industrial branches in which they originated or were produced, we pass to the distribution of national product by type of income. Finally, we distinguish between consumers' outlay and capital formation as the two components of the national product total, and discuss the composition of capital formation and of consumers' outlay.

II TOTAL NATIONAL PRODUCT

1 THE TOTALS AND THEIR CHANGES

THE gross and net national product for this country since 1919, evaluated at the current market prices of each year, are given in the first and third columns of Table 1 (see also Chart 1). To repeat the three definitions, net national product or national income represents the value of all commodities and services produced during the year reduced by the current market value of all commodities consumed in the process of production; or the sum of all current income payments to individuals plus the net savings of business and other enterprises; or the sum of market values of all commodities and services consumed by individuals, plus individuals' savings, plus net savings of business and other enterprises. Gross national product is defined similarly except that it includes also the value of durable capital goods consumed in the process of production, and correspondingly includes gross savings of enterprises rather than net.

Thus defined and measured in current market values, the national product of the United States has changed markedly since the World War. From a trough in 1921 of 66.1 billion dollars, which followed immediately a peak in 1920 of 82.8 billion, gross national product rose to the next peak, in 1929, of 93.6 billion. The contraction that followed

Table 1

NATIONAL INCOME AND GROSS NATIONAL PRODUCT, 1919–1935[1]

YEAR	GROSS NATIONAL PRODUCT		NET NATIONAL PRODUCT OR NATIONAL INCOME		NATIONAL INCOME IN 1929 PRICES PER		
	CURRENT PRICES	1929 PRICES	CURRENT PRICES	1929 PRICES	CAPITA	GAINFULLY OCCUPIED	CONSUMING UNIT
	(millions of dollars)		(millions of dollars)			(dollars)	
	(1)	(2)	(3)	(4)	(5)	(6)	(7)
1919	68,750	63,975	59,926	55,846	533	1,345	749
1920	82,836	66,888	72,386	58,759	553	1,405	778
1921	66,148	62,550	58,343	54,754	507	1,284	712
1922	67,186	68,482	59,706	60,310	549	1,395	771
1923	78,214	77,411	69,706	69,080	619	1,568	868
1924	78,791	78,272	70,369	69,868	615	1,552	861
1925	83,413	81,827	74,846	73,097	633	1,598	887
1926	88,780	86,363	79,477	76,939	657	1,657	919
1927	86,778	85,790	77,429	76,349	644	1,619	899
1928	90,053	90,168	80,397	80,352	669	1,679	933
1929	93,640	93,623	83,424	83,407	687	1,717	956
1930	82,723	84,872	72,940	74,646	608	1,513	844
1931	64,751	72,649	56,010	62,539	505	1,253	700
1932	47,202	58,256	39,628	48,560	389	961	537
1933	46,538	60,482	39,283	50,998	406	998	559
1934	55,765	68,924	47,849	59,272	468	1,146	644
1935	61,243	73,281	53,035	63,502	499	1,213	684
Average I 1919–1934	73,848	75,033	65,107	65,924	565	1,418	789
Average II 1919–1926	76,765	73,221	68,095	64,832	583	1,476	818
Average III 1927–1934	70,931	76,846	62,120	67,015	547	1,361	759
Percentage change from Average II to Average III	−7.6	+5.0	−8.8	+3.4	−6.2	−7.8	−7.2

[1] The estimates of gross and net national product for 1935 in this and subsequent tables rest upon a much more slender foundation than the measures for the earlier years (for the basis of the estimates for 1935 see the discussion in Appendix A). For this reason the averages for the period or parts of the period often exclude the figure for 1935.

The estimates of national income differ slightly from those shown in Table 1, *Bulletin 66* (September 27, 1937) because of minor revisions in the volume of net capital formation and a change in the technique of adjusting for price changes.

TOTAL NATIONAL PRODUCT

Chart 1
NATIONAL PRODUCT, 1919-1935

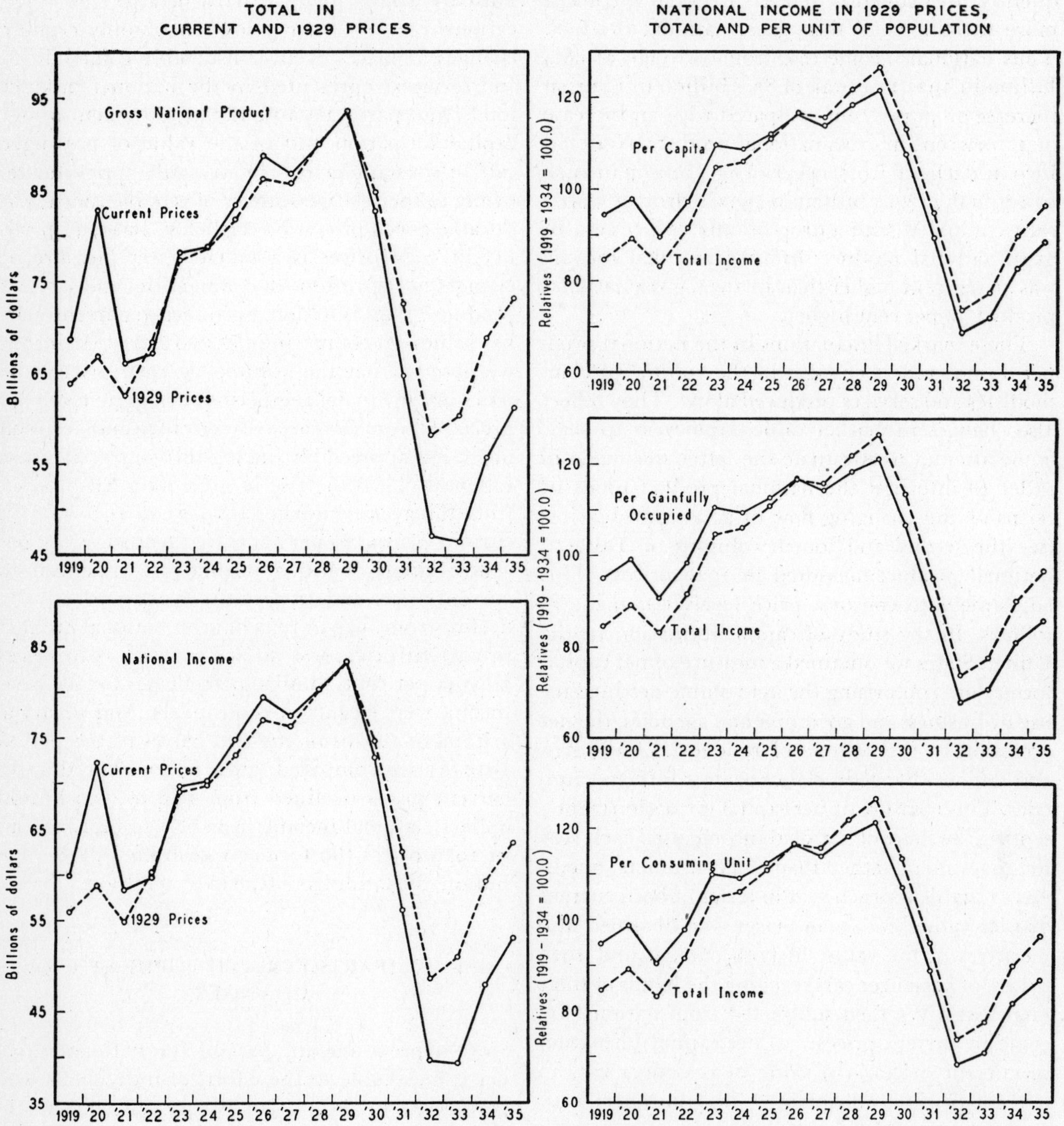

was severe; gross national product fell 50 per cent to a trough of 46.5 billion dollars in 1933. It has since risen materially, but in 1935 was still 35 per cent below the 1929 level.

Net national product or national income differs from gross national product in that it excludes the estimated consumption of durable capital goods. The annual volume of this consumption, which averaged about 9 billion dollars during the period, is measured largely by depreciation and depletion charges;[5] and the durable capital goods to which the charges refer include all capital equipment owned by business enterprises, whether corporate or individual, and by governmental agencies. They also include all real estate, whether residential or business, and whether owned by corporations or individuals.

The estimated consumption of durable capital

[5] Fire and marine losses are also included but the item is negligible compared with depreciation and depletion charges.

[9]

goods is far more stable than the sum of the other components of gross national product. Consequently, its deduction renders national income a more variable total than gross national product. Thus national income rose from a trough of 58.3 billion in 1921 to a peak of 83.4 billion in 1929, an increase of 43 per cent compared with an increase of 42 per cent in gross national product. National income declined from a peak of 83.4 billion in 1929 to a trough of 39.3 billion in 1933, a drop of 53 per cent compared with a drop of only 50 per cent in gross national product. In 1935 national income was 35 per cent higher than in 1933, gross national product 32 per cent higher.

These marked fluctuations in the national product do not reflect changes in the volume of commodities and services produced alone. They reflect also changes in market values from year to year. Some attempt to eliminate the latter was made in order to interpret the national product totals in terms of the changing flow of goods and services (see the second and fourth columns of Table 1, national product measured in 1929 prices). This adjustment to constant price levels was made as follows. In the study of capital formation in the United States we obtained a measure of net capital formation, comprising the net volume destined for use by business and governmental agencies, the net accretion to the volume of residential buildings, and the net change in claims against foreign countries. This measure of net capital formation, representing the share of national income used each year for investment rather than for consumers' goods, was available for each year in terms of both current market values and 1929 prices. We obtained also measures of the value, in both current and 1929 prices, of passenger cars reaching the hands of their final users. We then subtracted from national income in current prices: (a) net capital formation in current prices; (b) value of passenger cars in current prices; and obtained (c) consumers' outlay, exclusive of that for passenger cars. The residue under (c) was adjusted for changes in prices of consumption goods, on the assumption that they were reflected in the index of wage earners' cost of living compiled by the United States Bureau of Labor Statistics; then the adjusted residue (c^1) was recombined with (b^1) value of passenger cars in 1929 prices and (a^1) net capital formation in 1929 prices. The result was an estimate of national income in 1929 prices. The addition to it of the value of durable capital goods consumed in the process of production, measured in 1929 prices, yielded a measure of gross national product in 1929 prices.[6]

This adjustment for changes in price level is admittedly approximate, partly because the wage earners' cost of living index reflects only crudely changes in prices of all consumable commodities and services represented in the national product totals; and partly because the measurement of net capital formation and of the value of passenger cars in constant prices is necessarily approximate, owing to the relative scarcity of reliable indexes of durable goods prices. Nevertheless, national product in 1929 prices is a much better measure of changes in the volume of commodities and services produced than is national product in current market values. In comparing the two sets of estimates, we observe that the volume in 1929 prices rose much more during the period, i.e., grew more appreciably from the early years to 1929 and declined much less appreciably during the contraction that followed. Thus the rise in gross national product from the average during 1919–21 to 1929 was, in current prices, 29 per cent; in 1929 prices, 45 per cent. Similarly, national income rose 31 per cent in current prices and 48 per cent in 1929 prices. The decline from 1929 to 1933 in gross national product in current prices was 50 per cent; in 1929 prices only 35 per cent. Similar percentages for national income were 53 and 39, respectively. And when the arithmetic means of the two halves of the period 1919–34 are compared, gross national product in current prices declined from 76.8 to 70.9 billion dollars, national income from 68.1 to 62.1 billion; in 1929 prices the former rose from 73.2 to 76.8 billion, the latter from 64.8 to 67.0 billion.

2 COMPARISON WITH POPULATION CHANGES

Since a preponderant part of the national product is imputable to the efforts of individuals who participate in the productive process, and a very large part becomes available for consumption by the individuals of whom the nation is composed, we compare changes in the national product with

[6] The measures of the value in 1929 prices of durable capital goods consumed were provided by Mr. Fabricant, as were all the other measures of capital consumption. For the estimates of net capital formation see Section VIII below. The choice of 1929 prices as the constant price level in which to express national product was dictated by its corresponding use in the capital formation study. However, since the 1929 price level was only slightly below the 1926 and only slightly above the average price level for the entire period, 1919–35, it may be taken to represent, with respect to the level of market values, an average rather than an exceptional year.

changes in the number of persons to whom it is so closely related. This comparison is most significant when net national product or national income in constant prices is used, since national income offers a closer approximation than does gross national product to that part of the national product that is imputable to efforts of individuals; and since national income in constant prices offers a better approximation to the changing volume of commodities and services than does national income in fluctuating current prices.

In comparing changes in total income with those in the number of individuals, two questions may be raised: (a) what has been the net product per individual available for participation in the productive process? In other words, have the changes in the command over goods and services represented by national income been associated only with those in the number of individuals available for participation in the productive process, or also with other factors? (b) How have the changes in national income, representing the flow of goods and services available for consumption, compared with changes in the number of individuals who make up the community of consumers?

Both questions are answered roughly by a comparison of changes in national income with changes in total population and the computation of per capita income (Table 1, column 5). The rise in net national product from 1921 to 1929 and the decline after 1929 were accompanied by an increase in population; consequently per capita income rose appreciably less and declined more than did total income. Thus, while the latter rose from 54.8 billion dollars in 1921 to 83.4 billion in 1929, or 52 per cent, per capita income rose from $507 to $687, only 36 per cent. The declines from 1929 to 1932 were 42 and 43 per cent, respectively. The averages for the two halves of the post-War period show that when reduced to a per capita basis, the rise observed in total income becomes a decline.

However, changes in total population are but a crude approximation to the changes that may have occurred in the number of individuals available for direct participation in the productive process; or in the number of equivalent consuming units in the nation. Certain age and sex groups are, for obvious reasons, not capable of participating in production; and a child is not the equivalent of an adult with respect to consumption demands. We have, therefore, obtained estimates that measure more directly the number of producing and consuming units in the nation. To represent the former the number of people gainfully occupied was taken; for the latter we have utilized the age and sex distribution of the population for each year in the period to convert total population to the number of consuming units.[7]

The gainfully occupied group is composed of those members of the total population who ordinarily are engaged in pursuits whose product is included in national income or in gross national product. It thus covers not only persons who work for a monetary return but also farmers and other workers who may be compensated in kind, and it includes them whether or not they happen to be employed at the given point of time. The 'consuming unit' is the result of a statistical device by which the differences in needs for food, clothing, shelter, transportation, etc., of various age and sex groups in the population are roughly taken into account. Persons in certain age and sex groups, specifically adult males between twenty and thirty-four years old, are considered full bodied consumers, and each is taken to represent a complete consuming unit. The members of younger and older age groups, whose needs are naturally not as great as those of an adult male during the working period of his life, are each assigned a fractional value designed to indicate the magnitude of their needs as consumers as compared with that of the full consuming unit. Obviously, in determining the size of the gainfully occupied group or of the consuming unit equivalent of population, the adult groups, especially the male, have considerably greater weight than the age groups under eighteen.

Because of the growing proportion of adult groups in the total population, both the number gainfully occupied and the number of consuming units increased more rapidly than the total population; consequently, when national income is computed per gainfully occupied or per consuming unit, the movement in these per unit figures is algebraically smaller than in per capita income. Thus while per capita income rose from $507 in 1921 to $687 in 1929, or 36 per cent, both the income per person gainfully occupied and that per consuming unit rose 34 per cent; the declines from 1929 to 1932 were 43 per cent in per capita income; in the other two, 44 per cent. The differences be-

[7] For the annual estimates of the number of gainfully occupied we are indebted to Daniel Carson of the National Research Project on Reemployment Opportunities and Recent Changes in Industrial Techniques; for the estimates of the age and sex distribution of the population we are indebted to W. S. Thompson and P. K. Whelpton of the Scripps Foundation, Miami, Ohio. The consuming equivalents for each age and sex group are those given in their monograph, *Population Trends in the United States* (McGraw-Hill, 1933), p. 169.

tween the changes in the average per capita income for the two halves of the post-War period and the corresponding changes in the average income per gainfully occupied and per consuming unit are similarly small, but significant.

III DISTRIBUTION ACCORDING TO INDUSTRIAL ORIGIN

1 MEANING OF CLASSIFICATION

THE distribution of the national product according to industrial origin reveals in which industries the net and gross supply of commodities and services are produced. Tables 2 and 3 indicate, in dollar volumes and in percentages of the total, the amounts the various branches of the productive system contribute to gross and net national product. But the procedure by which these estimates have been obtained allows also the apportionment of income payments to individuals by the industrial characteristics of the enterprises making them. Accordingly, Table 4 gives the distribution of aggregate income payments to individuals by industrial sources.

The industrial divisions distinguished in Tables 2, 3 and 4 call for little explanation. Construction refers to contract construction alone, and does not include income originating in construction activities undertaken by business enterprises and public agencies on their own account. The transportation and other public utilities group includes throughout the following subdivisions: (a) electric light and power and manufactured gas; (b) steam railroads, Pullman and express; (c) other transportation, including pipe lines, street railways and water transportation; (d) communication, including telephone and telegraph.[8] Trade covers both the wholesale and retail branches of distribution. The finance group covers: (a) commercial banking; (b) insurance, both life and other; (c) real estate. Real estate includes, in addition to income arising from management and handling of real estate units by corporations especially engaged in that field, all net rents to individuals owning real estate, whether in cash received by individual owners from tenants or imputed to owners of non-farm residential units who reside on their property.[9]

Government covers not only the Federal government but also all other governmental units in the United States, including income originating in public education and the Post Office. Service covers the numerous branches of service activities: religious, professional, domestic, personal, recreation and amusement, and business. The miscellaneous group is a catch-all for the various activities that cannot be measured separately and properly under the relevant division. It includes such highly dissimilar enterprises as truck and bus transportation, taxicabs, and brokerage houses.

The estimates in Table 3 were obtained by adding to income payments to individuals made by enterprises in each industrial division the net savings of these enterprises. But to get an adequate measure of business savings the available data from accounting records on business profit and loss after payment of dividends must be adjusted in several ways; and some of the adjustments can be made only for the national income total as a whole or for the major industrial divisions.[10] Thus the correction for the difference between depreciation charges in cost and reproduction prices can be made only for national income as a whole, not for the various industrial divisions. For this reason, Table 3 includes an item of net business savings that is uncorrected for the disparity between depreciation and depletion charges at book value and at reproduction prices; and the total (line 11) differs in that respect from national income in current prices as shown in Table 1. However, this adjustment is relatively minor, and the effect of its omission on the distribution by industrial sources is insignificant.

The measures in Table 2 were obtained by adding to those in Table 3 the estimated volume of fixed capital consumption; and since most of the measures in Table 3 assume that depreciation and

[8] Estimates of national income and its elements originating in these subdivisions, as well as in the subdivisions of the finance group listed below, are given in Appendix Tables I and II.

[9] In the distribution by type of income rents appear as a separate type. But since rents cannot be apportioned by the industrial characteristics of the enterprises that pay them, they have to be treated in the industrial classification as entrepreneurial income payments in the real estate industry.

[10] For a more detailed discussion of these adjustments see Section IV.

depletion charges are based on book value rather than on current reproduction prices, the volume of capital consumption added to them to obtain the entries in Table 2 was also based upon the same assumption.[11] But since the estimates of capital consumption are based on data different from those underlying the measures of net income, the industrial divisions for the two sets of measures are not strictly comparable. However, sizable differences in scope are present for only three industrial divisions: transportation and other public utilities, finance, and miscellaneous. The measures of capital consumption for transportation and other public utilities cover not only the four subdivisions mentioned above but also such relatively minor fields as aerial transportation, bus lines, and cartage and storage (included in the national income classification under miscellaneous). Similarly under finance, the measures of capital consumption include, besides the three subdivisions noted above, such enterprises as stock and bond brokers, sales finance companies, and loan companies (included in the national income classification under miscellaneous). The effect is to exaggerate in Table 2 the values shown for the transportation and other public utilities and the finance groups, if they are understood, as they should be, to comprise the subdivisions indicated in Appendix Table I and noted at the beginning of this Section; and to underestimate the value shown for the division of miscellaneous industries. But the distortion is hardly significant, except for the miscellaneous division, which, in any case, has little independent value in the industrial classification. It would scarcely affect the broad conclusions revealed by the distribution according to industrial origin; and with reference to these broad conclusions the classifications in Tables 2, 3 and 4 can be considered as practically identical.

2 DISTRIBUTION IN CURRENT PRICES

The industrial divisions in Tables 2, 3 and 4 are assembled in three major groups. Group I, comprising agriculture, mining, manufacturing and construction, may be characterized as the industries

[11] For this reason, gross national product in Table 2 (line 11) and in Table 1 are equal. Gross national product would be the same, whether we add to national income, corrected for the disparity between depreciation and depletion charges based on book value and on current reproduction price, estimates of capital consumption that assume current reproduction price as the basis; or whether we add to national income, unadjusted for the disparity just mentioned, estimates of capital consumption that assume book value as the basis; see also Appendix D.

dealing primarily and largely with the production of commodities. Group II, comprising transportation and other public utilities and trade, may be characterized as commodity handling. This characterization, while true of trade, is only roughly true of transportation and other public utilities because even steam railroads carry passengers and the activities of such divisions as communication can hardly be classified as commodity handling. Still, the preponderant part of even the combined transportation and other public utilities group is devoted to commodity handling, rather than to commodity production or the provision of services to individuals. Group III, comprising finance, government, service and miscellaneous, may be characterized broadly as service industries; and while it includes some commodity handling under miscellaneous, the relative share of this activity or of commodity production is probably very small.

Table 3, Part B, reveals that a large share of total national income arises from activities that do not constitute either production or handling of new commodities; on the average, these account together for slightly over 61 per cent of national income. The rest is accounted for largely by activities that are either services by individuals to other individuals or to society as a whole, services rendered by highly durable commodities to individuals, or services rendered in connection with the production and handling of commodities but so distinct from them as to constitute a separate industry.

The industrial distribution of aggregate income payments to individuals (Table 4, Part B), when studied for the same three broad industrial groups, does not differ materially from that of national income. When an arithmetic mean of the percentage shares is taken for the period as a whole, the average share in the aggregate income payments to individuals is: Group I, 37 per cent; Group II, 24 per cent; Group III, 39 per cent. Similar average percentages for the percentage distribution of national income are 37, 25, and 39, respectively; the differences between the two sets of averages are insignificant. Nor does any significant difference appear when we consider the distribution of gross national product: the average share of Group I in Table 2, Part B, is 38 per cent; of Group II, 24 per cent; of Group III, 39 per cent.

As among the three groups, there were clear-cut differences in the movement over the period as a whole (see Chart 2). Whether expressed in percentages of the national product or of aggregate income payments, the relative share of the commod-

Table 2

DISTRIBUTION OF GROSS NATIONAL PRODUCT ACCORDING TO INDUSTRIAL ORIGIN, 1919-1934[1]

Part A. Absolute Figures

(millions of dollars)

	1919	1920	1921	1922	1923	1924	1925	1926	1927	1928	1929	1930	1931	1932	1933	1934
1 Agriculture	12,778	10,578	7,556	7,037	7,888	8,523	9,028	8,578	8,517	8,646	8,616	6,810	4,624	3,212	4,035	5,287
2 Mining	2,340	3,121	2,251	2,014	2,888	2,497	2,546	2,821	2,429	2,158	2,386	1,865	1,188	783	782	1,383
3 Manufacturing	17,322	21,200	13,914	14,565	18,357	17,150	18,482	20,054	19,147	19,989	21,984	18,373	12,972	8,049	8,389	11,372
4 Construction	1,564	2,342	1,975	1,985	2,690	2,974	3,051	3,230	3,240	3,213	3,224	2,785	1,816	816	592	874
5 Transportation and other public utilities	6,359	7,798	6,807	6,683	7,586	7,661	8,220	8,679	8,652	8,966	9,524	8,764	7,541	5,944	5,595	5,670
6 Trade	10,367	11,851	10,000	8,870	10,435	10,124	10,539	11,936	10,942	11,422	11,729	11,469	9,372	6,517	5,312	6,305
7 Finance	7,786	8,926	8,482	9,002	9,756	10,497	10,862	11,254	11,610	12,517	12,625	11,389	9,339	7,290	6,486	6,662
8 Government	1,368	7,427	6,571	6,961	7,603	7,826	8,108	8,658	9,085	9,080	9,137	8,791	7,239	6,407	7,602	8,440
9 Service	6,503	7,129	6,528	7,853	8,451	8,986	9,811	10,618	10,278	10,997	11,577	10,490	8,951	6,788	6,481	7,607
10 Miscellaneous	2,334	2,481	2,075	2,209	2,590	2,633	2,878	3,084	3,038	3,258	3,079	2,213	1,885	1,538	1,361	2,250
11 Total[2]	68,750	82,836	66,148	67,186	78,215	78,791	83,414	88,780	86,778	90,051	93,640	82,724	64,752	47,201	46,538	55,764
12 Group I (1-4)	34,004	37,241	25,696	25,600	31,824	31,144	33,107	34,684	33,334	34,006	36,209	29,833	20,600	12,860	13,798	18,915
13 Group II (5 and 6)	16,725	19,649	16,807	15,553	18,022	17,785	18,758	20,615	19,594	20,388	21,253	20,233	16,914	12,461	10,907	11,975
14 Group III (7-10)	17,991	25,964	23,656	26,025	28,400	29,942	31,660	33,615	34,011	35,852	36,418	32,883	27,415	22,022	21,932	24,959

[1] The grand totals in this and subsequent tables are additions of the totals of each industrial branch. The income subtotals by industrial branches are primarily in thousands of dollars. Hence the rounded figures given here for the individual industries do not always add up exactly to the totals shown.

[2] Inclusive of non-allocable items of capital consumption consisting of fire losses and depreciation on passenger cars used for business.

Part B. Percentage Distribution

	1919	1920	1921	1922	1923	1924	1925	1926	1927	1928	1929	1930	1931	1932	1933	1934
1 Agriculture	18.6	12.8	11.4	10.5	10.1	10.8	10.8	9.6	9.8	9.6	9.2	8.2	7.1	6.8	8.7	9.5
2 Mining	3.4	3.8	3.4	3.0	3.7	3.2	3.0	3.2	2.8	2.4	2.5	2.2	1.8	1.7	1.7	2.5
3 Manufacturing	25.2	25.6	21.0	21.7	23.5	21.7	22.1	22.6	22.0	22.1	23.4	22.1	20.0	17.0	18.0	20.4
4 Construction	2.3	2.8	3.0	3.0	3.4	3.8	3.7	3.6	3.7	3.6	3.4	3.4	2.8	1.7	1.3	1.6
5 Transportation and other public utilities	9.3	9.4	10.3	9.9	9.7	9.7	9.8	9.8	10.0	9.9	10.1	10.6	11.6	12.6	12.0	10.2
6 Trade	15.1	14.3	15.1	13.2	13.3	12.8	12.6	13.4	12.6	12.7	12.5	13.8	14.4	13.8	11.4	11.3
7 Finance	11.3	10.8	12.8	13.4	12.5	13.3	13.0	12.7	13.4	13.9	13.4	13.7	14.4	15.4	13.9	11.9
8 Government	2.0	9.0	9.9	10.4	9.7	9.9	9.7	9.7	10.4	10.1	9.7	10.6	11.1	13.5	16.3	15.1
9 Service	9.5	8.6	9.9	11.7	10.8	11.4	11.7	11.9	11.8	12.2	12.3	12.6	13.8	14.3	13.9	13.6
10 Miscellaneous	3.4	3.0	3.1	3.3	3.3	3.3	3.4	3.5	3.5	3.6	3.3	2.7	2.9	3.2	2.9	4.0
11 Total[1]	100.0	100.0	100.0	100.0	100.0	100.0	100.0	100.0	100.0	100.0	100.0	100.0	100.0	100.0	100.0	100.0
12 Group I (1-4)	49.5	44.9	38.8	38.1	40.7	39.5	39.6	39.0	38.3	37.7	38.6	36.0	31.7	27.2	29.6	33.9
13 Group II (5 and 6)	24.3	23.7	25.4	23.2	23.0	22.5	22.5	23.2	22.5	22.6	22.6	24.4	26.1	26.3	23.4	21.4
14 Group III (7-10)	26.2	31.3	35.8	38.7	36.3	38.0	37.9	37.8	39.1	39.7	38.8	39.6	42.2	46.5	47.0	44.7

[1] Exclusive of the non-allocable items which range from less than 0.1 per cent of -0.3 per cent of the totals.

Table 3

DISTRIBUTION OF NATIONAL INCOME ACCORDING TO INDUSTRIAL ORIGIN, 1919-1935[1]

Part A. Absolute Figures

(millions of dollars)

	1919	1920	1921	1922	1923	1924	1925	1926	1927	1928	1929	1930	1931	1932	1933	1934	1935
1 Agriculture	11,236	8,986	6,226	5,828	6,690	7,325	7,769	7,294	7,199	7,282	7,231	5,510	3,488	2,182	3,080	4,276	4,696
2 Mining	1,832	2,508	1,709	1,430	2,183	1,810	1,954	2,288	1,945	1,719	1,893	1,471	872	503	503	1,007	1,116
3 Manufacturing	16,171	19,907	12,627	13,079	16,784	15,598	16,833	18,151	17,186	17,927	19,828	16,179	10,997	6,248	6,644	9,814	11,546
4 Construction	1,501	2,258	1,895	1,910	2,617	2,892	2,955	3,128	3,129	3,100	3,091	2,646	1,688	708	506	795	878
5 Transportation and other public utilities	5,989	7,461	6,371	6,228	7,089	7,127	7,624	7,934	7,838	8,043	8,513	7,715	6,395	4,860	4,583	4,654	4,991
6 Trade	10,115	11,587	9,692	8,562	10,091	9,742	10,082	11,486	10,465	10,895	11,177	10,901	8,789	6,006	4,846	5,819	6,863
7 Finance	5,397	5,918	6,299	6,810	7,363	7,997	8,304	8,489	8,786	9,551	9,541	8,344	6,644	4,906	4,171	4,193	4,643
8 Government	923	6,906	6,151	6,555	7,153	7,361	7,628	8,144	8,531	8,505	8,535	8,184	6,641	5,859	7,020	7,793	7,719
9 Service	6,413	7,022	6,408	7,691	8,290	8,798	9,611	10,195	10,015	10,693	11,224	10,136	8,602	6,375	6,111	7,236	8,261
10 Miscellaneous	2,264	2,416	2,014	2,159	2,538	2,607	2,861	3,083	3,034	3,257	3,077	2,211	1,883	1,537	1,361	2,248	2,397
11 Total	61,842	74,969	59,393	60,254	70,799	71,257	75,621	80,192	78,128	80,970	84,111	73,297	56,000	39,184	38,824	47,834	53,110
12 Group I (1-4)	30,740	33,659	22,457	22,248	28,274	27,624	29,510	30,861	29,459	30,027	32,044	25,805	17,045	9,641	10,733	15,891	18,236
13 Group II (5 and 6)	16,104	19,048	16,063	14,791	17,181	16,869	17,706	19,420	18,303	18,938	19,690	18,616	15,184	10,866	9,430	10,473	11,854
14 Group III (7-10)	14,998	22,262	20,872	23,215	25,344	26,764	28,405	29,911	30,367	32,005	32,377	28,875	23,770	18,677	18,662	21,469	23,020

[1] Not adjusted for the disparity between depreciation and depletion at cost prices and at current reproduction prices.

Part B. Percentage Distribution

	1919	1920	1921	1922	1923	1924	1925	1926	1927	1928	1929	1930	1931	1932	1933	1934	1935
1 Agriculture	18.2	12.0	10.5	9.7	9.4	10.3	10.3	9.1	9.2	9.0	8.6	7.5	6.2	5.6	7.9	8.9	8.8
2 Mining	3.0	3.3	2.9	2.4	3.1	2.5	2.6	2.9	2.5	2.1	2.3	2.0	1.6	1.3	1.3	2.1	2.1
3 Manufacturing	26.1	26.6	21.3	21.7	23.7	21.9	22.3	22.6	22.0	22.1	23.6	22.1	19.6	15.9	17.1	20.5	21.7
4 Construction	2.4	3.0	3.2	3.2	3.7	4.1	3.9	3.9	4.0	3.8	3.7	3.6	3.0	1.8	1.3	1.7	1.7
5 Transportation and other public utilities	9.7	10.0	10.7	10.3	10.0	10.0	10.1	9.9	10.0	9.9	10.1	10.5	11.4	12.4	11.8	9.7	9.4
6 Trade	16.4	15.5	16.3	14.2	14.3	13.7	13.3	14.3	13.4	13.5	13.3	14.9	15.7	15.3	12.5	12.2	12.9
7 Finance	8.7	7.9	10.6	11.3	10.4	11.2	11.0	10.6	11.2	11.8	11.3	11.4	11.9	12.5	10.7	8.8	8.7
8 Government	1.5	9.2	10.4	10.9	10.1	10.3	10.1	10.2	10.9	10.5	10.1	11.2	11.9	15.0	18.1	16.3	14.5
9 Service	10.4	9.4	10.8	12.8	11.7	12.3	12.7	12.7	12.8	13.2	13.3	13.8	15.4	16.3	15.7	15.1	15.6
10 Miscellaneous	3.7	3.2	3.4	3.6	3.6	3.7	3.8	3.8	3.9	4.0	3.7	3.0	3.4	3.9	3.5	4.7	4.5
11 Total	100.0	100.0	100.0	100.0	100.0	100.0	100.0	100.0	100.0	100.0	100.0	100.0	100.0	100.0	100.0	100.0	100.0
12 Group I (1–4)	49.7	44.9	37.8	36.9	39.9	38.8	39.0	38.5	37.7	37.1	38.1	35.2	30.4	24.6	27.6	33.2	34.3
13 Group II (5 and 6)	26.0	25.4	27.0	24.5	24.3	23.7	23.4	24.2	23.4	23.4	23.4	25.4	27.1	27.7	24.3	21.9	22.3
14 Group III (7–10)	24.3	29.7	35.1	38.5	35.8	37.6	37.6	37.3	38.9	39.5	38.5	39.4	42.4	47.7	48.1	44.9	43.3

Table 4

DISTRIBUTION OF AGGREGATE INCOME PAYMENTS TO INDIVIDUALS ACCORDING TO INDUSTRIAL ORIGIN, 1919-1935

Part A. Absolute Figures

(millions of dollars)

	1919	1920	1921	1922	1923	1924	1925	1926	1927	1928	1929	1930	1931	1932	1933	1934	1935
1 Agriculture	8,480	9,757	6,592	6,350	7,022	7,094	7,197	7,351	7,333	7,328	7,354	6,617	5,258	4,041	3,813	4,152	4,545
2 Mining	1,796	2,494	1,861	1,703	2,428	2,099	2,054	2,382	2,111	1,886	2,059	1,712	1,186	799	790	1,111	1,167
3 Manufacturing	14,292	16,780	11,732	12,255	15,262	14,553	15,383	16,168	16,260	16,893	18,035	15,864	12,288	8,567	8,516	10,509	11,710
4 Construction	1,541	2,093	1,623	1,975	2,657	2,702	2,852	2,951	2,909	3,063	3,031	2,518	1,644	833	794	938	908
5 Transportation and other public utilities	5,867	7,287	6,059	6,014	6,690	6,751	6,956	7,243	7,378	7,405	7,777	7,498	6,554	5,306	4,803	4,964	5,205
6 Trade	8,034	9,039	7,676	8,032	8,802	9,029	9,478	9,970	9,807	10,030	10,650	10,178	8,926	7,016	6,239	6,749	7,117
7 Finance	5,193	5,874	6,223	6,806	7,390	7,939	8,232	8,375	8,626	9,208	9,698	8,714	7,280	5,653	4,771	4,528	4,780
8 Government	5,202	5,299	5,492	5,553	5,728	5,896	6,058	6,267	6,514	6,741	7,028	7,209	7,341	7,265	7,838	8,974	9,364
9 Service	5,202	6,302	6,050	7,250	7,502	8,191	8,953	9,241	9,495	10,097	10,751	9,975	9,092	7,806	8,115	8,130	9,023
10 Miscellaneous	1,893	2,130	1,868	2,103	2,375	2,508	2,759	2,874	2,948	3,173	3,425	3,334	2,996	2,499	2,200	2,330	2,468
11 Total	57,499	67,056	55,177	58,041	65,854	66,763	69,921	72,823	73,381	75,823	79,808	73,620	62,565	49,785	47,880	52,385	56,287
12 Group I (1-4)	26,108	31,124	21,809	22,283	27,368	26,448	27,486	28,853	28,613	29,170	30,479	26,711	20,376	14,240	13,913	16,710	18,330
13 Group II (5 and 6)	13,901	16,326	13,735	14,046	15,492	15,780	16,433	17,213	17,185	17,434	18,427	17,676	15,480	12,322	11,042	11,713	12,322
14 Group III (7-10)	17,489	19,605	19,633	21,712	22,994	24,534	26,002	26,757	27,583	29,218	30,902	29,233	26,710	23,223	22,924	23,962	25,635

[18]

Part B. Percentage Distribution

	1919	1920	1921	1922	1923	1924	1925	1926	1927	1928	1929	1930	1931	1932	1933	1934	1935
1 Agriculture	14.7	14.6	11.9	10.9	10.7	10.6	10.3	10.1	10.0	9.7	9.2	9.0	8.4	8.1	8.0	7.9	8.1
2 Mining	3.1	3.7	3.4	2.9	3.7	3.1	2.9	3.3	2.9	2.5	2.6	2.3	1.9	1.6	1.6	2.1	2.1
3 Manufacturing	24.9	25.0	21.3	21.1	23.2	21.8	22.0	22.2	22.2	22.3	22.6	21.5	19.6	17.2	17.8	20.1	20.8
4 Construction	2.7	3.1	2.9	3.4	4.0	4.0	4.1	4.1	4.0	4.0	3.8	3.4	2.6	1.7	1.7	1.8	1.6
5 Transportation and other public utilities	10.2	10.9	11.0	10.4	10.2	10.1	9.9	9.9	10.1	9.8	9.7	10.2	10.5	10.7	10.0	9.5	9.2
6 Trade	14.0	13.5	13.9	13.8	13.4	13.5	13.6	13.7	13.4	13.2	13.3	13.8	14.3	14.1	13.0	12.9	12.6
7 Finance	9.0	8.8	11.3	11.7	11.2	11.9	11.8	11.5	11.8	12.1	12.2	11.8	11.6	11.4	10.0	8.6	8.5
8 Government	9.0	7.9	10.0	9.6	8.7	8.8	8.7	8.6	8.9	8.9	8.8	9.8	11.7	14.6	16.4	17.1	16.6
9 Service	9.0	9.4	11.0	12.5	11.4	12.3	12.8	12.7	12.9	13.3	13.5	13.5	14.5	15.7	16.9	15.5	16.0
10 Miscellaneous	3.3	3.2	3.4	3.6	3.6	3.8	3.9	3.9	4.0	4.2	4.3	4.5	4.8	5.0	4.6	4.4	4.4
11 Total	100.0	100.0	100.0	100.0	100.0	100.0	100.0	100.0	100.0	100.0	100.0	100.0	100.0	100.0	100.0	100.0	100.0
12 Group I (1-4)	45.4	46.4	39.5	38.4	41.6	39.6	39.3	39.6	39.0	38.5	38.2	36.3	32.6	28.6	29.1	31.9	32.6
13 Group II (5 and 6)	24.2	24.3	24.9	24.2	23.5	23.6	23.5	23.6	23.4	23.0	23.1	24.0	24.7	24.8	23.1	22.4	21.9
14 Group III (7-10)	30.4	29.2	35.6	37.4	34.9	36.7	37.2	36.7	37.6	38.5	38.7	39.7	42.7	46.6	47.9	45.7	45.6

ity producing branches declined. Most of the decline occurred from 1919 to 1921 and from 1929 to 1932, but there was a downward drift even from 1923 to 1929; and inspection of the percentages for each industrial division in this group shows that in three—agriculture, mining, and manufacturing

Chart 2
PERCENTAGE DISTRIBUTION OF GROSS NATIONAL PRODUCT, NATIONAL INCOME AND AGGREGATE INCOME PAYMENTS BY MAJOR INDUSTRIAL DIVISIONS
1919 – 1935

■ Commodity producing industries
▨ Commodity handling industries
☐ Service industries

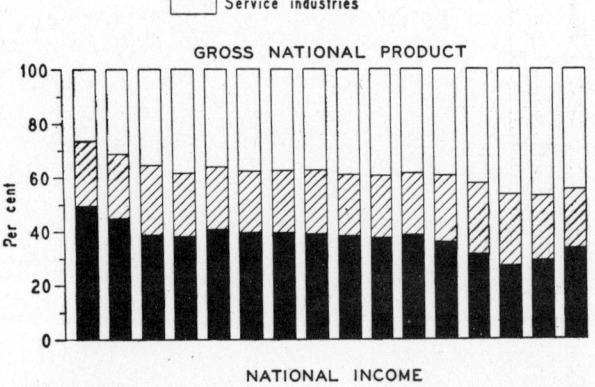
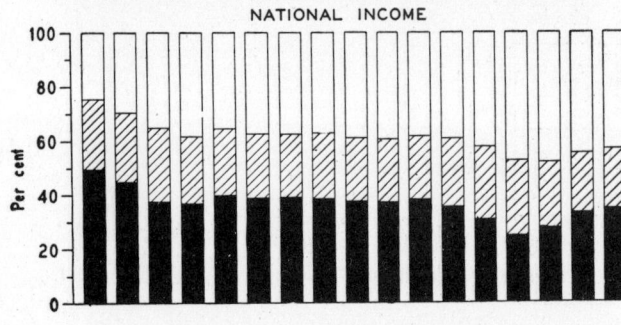
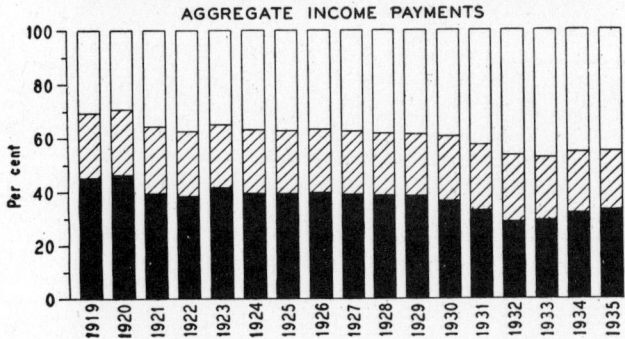

—this downward movement in the percentage share of gross and net national product and of aggregate income payments is clearly and consistently present. Only in construction, owing to the long swing characteristic of the industry, is this declining trend in relative importance not obvious. The share of the national product or of aggregate income payments to individuals accounted for by 'commodity handling', on the contrary, does not decline. It varies somewhat from year to year, both for the group as a whole and for its two divisions, but reveals no definite trend over the period. The distinct and different trends in the shares of the various industrial branches included under transportation and other public utilities over the period (e.g., the growth in the importance of electric light and power and gas, and of communications, and the decline in that of steam railroads and other transportation) are canceled when the branches are combined.

The share of each important subdivision of the service industries—finance, government, and service—rose distinctly over the period. The conclusion is thus unavoidable that the development of the economic system since the War has meant, as far as the composition of the national product in current prices is concerned, a shift from the production of commodities to service activities that are significantly different in character and industrial organization from the production and handling of commodities.

3 DISTRIBUTION IN 1929 PRICES

Are the changes in the industrial distribution of the totals in Tables 2–4 due to the differential movement of prices for various groups of commodities and services, or would they persist even with the dollar volumes adjusted for changes in the specific price levels? While data are scanty, some attempt to adjust for price changes can be made with reference to the broad industrial distribution of gross national product and of national income.

In *Bulletin 59* (May 4, 1936) we presented indexes of physical volume of output in several industrial branches, and indicated under what assumptions they were a good approximation to the changes in income produced in these branches in terms of a constant price level (see Appendix Table 2, p. 24, and discussion on p. 5). These assumptions can easily be modified to apply to gross income originating in industrial branches, 'gross' in the sense used in the present discussion.

Bulletin 59 provides indexes of the movement of net and gross income, in constant prices, originating in each of the four divisions of Group I (agriculture, mining, manufacturing, and construction). If these indexes, in terms of the level for 1919–34 as 100, are converted to the 1929 level as 100 and multiplied: (a) by income originating in each of the four divisions in Group I in 1929 (Table 3), and then added, the result would be annual estimates of the part of national income, in 1929 prices, originating in commodity producing industries; (b) by gross income originating in each of the four divisions in Group I in 1929 (Table 2), and then added, the result would be annual estimates of the part of gross national product in 1929 prices originating in the commodity producing industries. These results are entered in Table 5. We have also,

Table 5

DISTRIBUTION OF GROSS NATIONAL PRODUCT AND NATIONAL INCOME IN 1929 AND CURRENT PRICES BY ORIGIN IN COMMODITY PRODUCING AND IN OTHER INDUSTRIES, 1919–1934

Part A Gross National Product

| | ABSOLUTE TOTALS IN 1929 PRICES PRODUCT ORIGINATING IN | | PERCENTAGE DISTRIBUTION | | | |
| | | | 1929 PRICES | | CURRENT PRICES | |
YEAR	COMMODITY PRODUCING INDUSTRIES *(millions of dollars)* (1)	OTHER INDUSTRIES (2)	COMMODITY PRODUCING INDUSTRIES (3)	OTHER INDUSTRIES (4)	COMMODITY PRODUCING INDUSTRIES [1] (5)	OTHER INDUSTRIES [1] (6)
1919	25,589	38,386	40.0	60.0	49.5	50.5
1920	26,822	40,065	40.1	59.9	45.0	55.0
1921	22,759	39,792	36.4	63.6	38.8	61.2
1922	27,282	41,200	39.8	60.2	38.1	61.9
1923	30,270	47,141	39.1	60.9	40.7	59.3
1924	29,534	48,738	37.7	62.3	39.5	60.5
1925	31,589	50,238	38.6	61.4	39.7	60.3
1926	33,099	53,263	38.3	61.7	39.1	60.9
1927	33,030	52,760	38.5	61.5	38.4	61.6
1928	35,182	54,986	39.0	61.0	37.8	62.2
1929	36,209	57,414	38.7	61.3	38.7	61.3
1930	31,991	52,881	37.7	62.3	36.1	63.9
1931	28,887	43,762	39.8	60.2	31.8	68.2
1932	23,277	34,979	40.0	60.0	27.2	72.8
1933	24,483	35,999	40.5	59.5	29.6	70.4
1934	25,322	43,602	36.7	63.3	33.9	66.1
Average 1919–26			38.8	61.2	41.3	58.7
Average 1927–34			38.9	61.1	34.2	65.8

Part B National Income

1919	22,560	33,285	40.4	59.6	51.3	48.7
1920	23,621	35,137	40.2	59.8	46.5	53.5
1921	20,048	34,706	36.6	63.4	38.5	61.5
1922	24,104	36,206	40.0	60.0	37.3	62.7
1923	26,731	42,349	38.7	61.3	40.6	59.4
1924	26,069	43,798	37.3	62.7	39.3	60.7
1925	27,933	45,164	38.2	61.8	39.4	60.6
1926	29,252	47,687	38.0	62.0	38.8	61.2
1927	29,209	47,140	38.3	61.7	38.0	62.0
1928	31,137	49,215	38.8	61.2	37.3	62.7
1929	32,044	51,363	38.4	61.6	38.4	61.6
1930	28,238	46,408	37.8	62.2	35.4	64.6
1931	25,403	37,136	40.6	59.4	30.4	69.6
1932	20,359	28,201	41.9	58.1	24.3	75.7
1933	21,441	29,557	42.0	58.0	27.3	72.7
1934	22,216	37,056	37.5	62.5	33.2	66.8
Average 1919–26			38.7	61.3	41.5	58.5
Average 1927–34			39.4	60.6	33.0	67.0

[1] The percentages in Part A differ from those in Table 2, Part B, because of the inclusion here of non-allocable items with 'other' industries. The percentages in Part B differ from those in Table 3, Part B, because total national income distributed here, in contrast to that in Table 3, has been adjusted for the disparity between depreciation and depletion deductions in cost prices and in current reproduction prices.

from Table 1, the annual volume of gross and net national product in 1929 prices. By subtracting (b), i.e., the part originating in the commodity producing industries, from total gross national product in 1929 prices, we obtain the part in 1929 prices originating in all the other industries. Also, by subtracting (a), i.e., the part originating in the commodity producing industries, from total net national product or national income in 1929 prices, we obtain the part in 1929 prices originating in all the other industries. The latter subtraction involves some inconsistency since the national income figure in Table 1 has been adjusted for the disparity between depreciation and depletion charges on a book value and a current reproduction price basis, an adjustment that could not be made for the industrial divisions of the national total. But the adjustment is small and could scarcely affect the comparison between net income originating in the commodity producing industries and national income, both in constant prices. Accordingly, Table 5 presents this rough apportionment of gross and net national product into two major industrial groups: commodity producing industries and all other industries (see also Chart 3).

This adjustment for price changes is admittedly rough. While the adjusted estimates of gross and net national product, and of the parts originating in the commodity producing industries, may themselves be tolerably reliable, the derivation of the shares, in 1929 prices, of 'other industries' by subtraction makes the latter estimates bear the brunt of errors in both the total and the subtrahend. But the primary emphasis here is on the broad movement of the percentage distribution; and this movement is, perhaps, not too greatly affected by the crudities of the procedure.

It appears from Table 5 and Chart 3 that when the national product totals and those shares that originate in the commodity producing and other industries are adjusted for changes in prices, the tendency for the share of the former to decline disappears completely. If the estimates underlying Table 5 are to be trusted, the apparent decline over the period in the relative share of the commodity producing industries in the economic system, as measured by the industrial distribution of the national product, was due exclusively to the decline in the prices of commodities which was greater than in the prices of services. This conclusion checks with general impressions concerning the price movements of various categories of economic goods since the War. There is little doubt that ag-

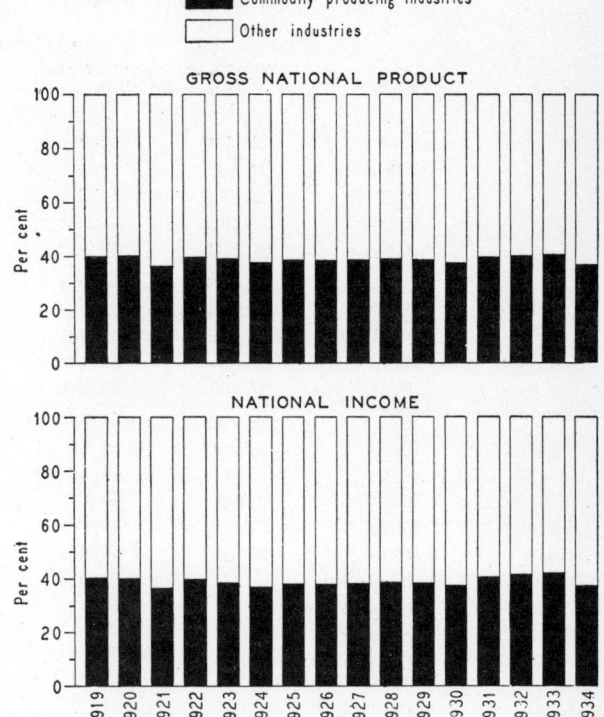

Chart 3
PERCENTAGE APPORTIONMENT OF GROSS NATIONAL PRODUCT AND NATIONAL INCOME IN 1929 PRICES, BY ORIGIN IN COMMODITY PRODUCING AND IN OTHER INDUSTRIES
1919 – 1934

ricultural products and other basic commodities declined more appreciably in price than transportation, distributive, financial and other services. That this difference in the price change was so substantial as to create the downward movement of the share in the national product, in current prices, of the commodity producing industries is indicated by Tables 2, 3 and 5. But in view of the slender basis upon which the difficult adjustment for price changes rests, it is perhaps best to leave this conclusion as a tentative suggestion deserving further exploration.

IV DISTRIBUTION ACCORDING TO TYPE OF INCOME

1 MEANING OF CLASSIFICATION

THE allocations discussed in this section have already been suggested in describing the composition of the national product totals. Business and other enterprises engaged in the production of these totals distribute most of their share as payments to individuals in return for their services or the services of their property; or in the form of payments to individuals only remotely connected with the productive process (pensions, compensation for injuries, relief payments, etc.). The residue, either positive or negative, represents in gross national product, the gross savings; and in national income, the net savings of these enterprises. It is of obvious importance to study how the national product is apportioned between such savings and aggregate income payments to individuals; and how the latter are apportioned among various types of payment. Such allocations can be made for the national product only as measured in current market values (Table 6).

Wages, salaries, pensions, dividend and interest payments are familiar. Dividend payments as measured here cover disbursements to individuals, and thus exclude the intercorporate dividend flow; interest payments cover interest on long term debt alone, upon the assumption that all short term interest is paid to other business enterprises and that short term interest paid to individuals by banks represents an indirect flow of long term interest; the only allowance for all industries for intercorporate long term interest payments is interest on government securities.[12] Entrepreneurial withdrawals are the amounts that individual entrepreneurs withdraw from their business for consumption and investment elsewhere—an estimate necessarily highly tentative. Rents received by individuals comprise both actual monetary rental received, minus the expenses incurred, and net rental imputed to the individuals who own the houses in which they reside (excluding owned farm homes). While compensation for injuries covers small amounts paid to individuals other than employees, it may be grouped with pensions, relief, wages and salaries as employees' compensation. Withdrawals by entrepreneurs and rents may be classified as entrepreneurial income payments, although there is some ground for classifying rent with property income. Dividends and interest receipts are individuals' property income.

Net savings of enterprises is a residual item of controlling importance, and should, in accordance with our definitions of national income, equal the difference between the current value of all commodities and services produced, less the current market value of commodities consumed in production, and total income payments to individuals. The entry on the income account of business enterprises that most closely corresponds to net savings as thus defined is net profit or loss, after deduction of dividends and taxes. But this item still differs from the desired measure of net savings in the following respects: (a) it includes gains and losses on the sale of capital assets; (b) it estimates the consumption of durable commodities, subject to depreciation and depletion, at their original cost, not at current reproduction cost; (c) it measures the consumption of current materials at original cost when prices are rising, or at cost or market whichever lower, when prices are declining, instead of evaluating them at the market price at the time of consumption.[13] Thus, in order to arrive at a correct figure for net savings of enterprises, the item available from business accounting records must be adjusted for all these possible departures, an adjustment that can be made in varying degree for the various distortions noted. The measurement of gross savings of enterprises, as already indicated, does not call for adjustment (b) but it does demand adjustments (a) and (c).

Gains and losses on the sale of capital assets can be excluded only since 1929; and savings, and hence national product, whether gross or net, are uncorrected in this respect for all years before 1929. The magnitude of this item can be judged by its amount, 1.6 billion dollars for 1932, the year

---

[12] These various assumptions are made largely because we lack data upon which to base a more accurate measure of interest payments to individuals. For most of the public utilities (electric light and power, steam railroads, Pullman and express, pipe lines, street railways and communications) the available data permit the deduction of interest received on all investments in both government and industrial securities.

[13] For a more detailed discussion of adjustments (b) and (c) see Parts Three and Four of *Studies in Income and Wealth*, Volume I, by the Conference on Research in National Income and Wealth (National Bureau of Economic Research, 1937); see also Appendix D below.

Table 6

DISTRIBUTION OF GROSS NATIONAL PRODUCT AND NATIONAL INCOME ACCORDING TO TYPE OF INCOME, 1919-1935

Part A. Absolute Figures

(millions of dollars)

		1919	1920	1921	1922	1923	1924	1925	1926	1927	1928	1929	1930	1931	1932	1933	1934	1935
1	Wages and salaries[1]	38,821	45,750	36,705	39,124	45,183	45,420	47,651	50,362	50,570	52,196	55,191	50,332	42,943	33,894	32,432	35,785	38,755
2	Compensation for injuries[2], pensions and relief[3]	476	636	720	730	755	779	723	718	749	768	809	851	953	976	1,920	2,764	2,876
3	Employees' compensation[1,2,3]	39,297	46,387	37,424	39,854	45,938	46,199	48,373	51,080	51,319	52,964	56,000	51,182	43,896	34,870	34,353	38,549	41,631
4	Withdrawals by entrepreneurs[4]	9,651	11,095	7,993	7,751	8,271	8,455	8,593	8,544	8,499	8,538	8,677	7,910	6,554	5,303	5,062	5,515	5,994
5	Rents	2,670	2,999	3,296	3,754	3,915	4,169	4,146	3,902	3,675	3,788	3,763	3,111	2,088	1,319	1,311	1,165	1,323
6	Entrepreneurial income payments	12,321	14,094	11,289	11,505	12,186	12,624	12,739	12,446	12,174	12,327	12,440	11,021	8,642	6,622	6,373	6,680	7,317
7	Dividends[5]	2,812	3,123	2,848	2,920	3,714	3,655	4,237	4,601	4,892	5,209	5,807	5,708	4,270	2,679	2,147	2,497	2,845
8	Interest[5]	3,041	3,410	3,546	3,676	3,922	4,156	4,418	4,586	4,864	5,174	5,409	5,507	5,474	5,362	4,837	4,566	4,445
9	Property income payments[6]	5,882	6,575	6,463	6,682	7,730	7,940	8,808	9,297	9,888	10,532	11,367	11,417	10,028	8,293	7,154	7,156	7,339
10	Aggregate income payments to individuals	57,499	67,056	55,177	58,041	65,854	66,763	69,921	72,823	73,381	75,823	79,808	73,620	62,565	49,785	47,880	52,385	56,287
11	Net savings of business enterprises[7]	6,706	3,722	2,506	663	2,427	2,141	3,355	4,777	2,031	2,810	2,109	-1,656	-5,856	-8,751	-7,779	-3,355	-1,607
12	Net savings of government	-4,279	1,607	660	1,002	1,425	1,465	1,571	1,877	2,017	1,764	1,507	975	-700	-1,406	-818	-1,181	-1,645
13	Total net savings of enterprises	2,427	5,330	3,166	1,665	3,853	3,606	4,926	6,654	4,048	4,574	3,616	-680	-6,556	-10,157	-8,596	-4,536	-3,252
14	National income	59,926	72,386	58,343	59,706	69,706	70,369	74,846	79,477	77,429	80,397	83,424	72,940	56,010	39,628	39,283	47,849	53,035
15	Gross savings of enterprises	11,251	15,780	10,971	9,145	12,361	12,028	13,493	15,957	13,397	14,230	13,832	9,103	2,186	-2,583	-1,341	3,380	4,956
16	Gross national product	68,750	82,836	66,148	67,186	78,214	78,791	83,413	88,780	86,778	90,053	93,640	82,723	64,751	47,202	46,538	55,765	61,243

[1] Including withdrawals by entrepreneurs in service and miscellaneous.
[2] Including payments by steam railroads, Pullman and express for compensation for injuries to persons other than employees.
[3] Including payment by the government for direct relief and work relief in 1933, 1934 and 1935.
[4] Excluding withdrawals by entrepreneurs in service and miscellaneous which are included with wages and salaries.
[5] A small amount of dividends from corporations engaged in agriculture is included with interest payments.
[6] Including net balance of international payments.
[7] These net savings are adjusted throughout for the effects of changing inventory valuations and for the difference between depreciation and depletion at cost prices and at current reproduction prices. But the adjustment for gains and losses on the same of capital assets was possible only since 1929. The items for these years, the adjustment for which is already made in the table, are as follows (in millions of dollars):

1929	1930	1931	1932	1933	1934	1935
816	-337	-1,509	-1,621	-1,500	-376	-157

[24]

Part B. Percentage Distribution

	1919	1920	1921	1922	1923	1924	1925	1926	1927	1928	1929	1930	1931	1932	1933	1934	1935
1 Wages and salaries	67.5	68.2	66.5	67.4	68.6	68.0	68.1	69.2	68.9	68.8	69.2	68.4	68.6	68.1	67.7	68.3	68.9
2 Compensation for injuries, pensions and relief	0.8	0.9	1.3	1.3	1.1	1.2	1.0	1.0	1.0	1.0	1.0	1.2	1.5	2.0	4.0	5.3	5.1
3 Employees' compensation	68.3	69.2	67.8	68.7	69.8	69.2	69.2	70.1	69.9	69.9	70.2	69.5	70.2	70.0	71.7	73.6	74.0
4 Withdrawals by entrepreneurs	16.8	16.5	14.5	13.4	12.6	12.7	12.3	11.7	11.6	11.3	10.9	10.7	10.5	10.7	10.6	10.5	10.6
5 Rents	4.6	4.5	6.0	6.5	5.9	6.2	5.9	5.4	5.0	5.0	4.7	4.2	3.3	2.6	2.7	2.2	2.4
6 Entrepreneurial income payments	21.4	21.0	20.5	19.8	18.5	18.9	18.2	17.1	16.6	16.3	15.6	15.0	13.8	13.3	13.3	12.8	13.0
7 Dividends	4.9	4.7	5.2	5.0	5.6	5.5	6.1	6.3	6.7	6.9	7.3	7.8	6.8	5.4	4.5	4.8	5.1
8 Interest	5.3	5.1	6.4	6.3	6.0	6.2	6.3	6.3	6.6	6.8	6.8	7.5	8.7	10.8	10.1	8.7	7.9
9 Property income payments	10.2	9.8	11.7	11.5	11.7	11.9	12.6	12.8	13.5	13.9	14.2	15.5	16.0	16.7	14.9	13.7	13.0
10 Aggregate income payments to individuals	100.0	100.0	100.0	100.0	100.0	100.0	100.0	100.0	100.0	100.0	100.0	100.0	100.0	100.0	100.0	100.0	100.0
11 Total net savings of enterprises as percentage of aggregate income payments to individuals	4.2	7.9	5.7	2.9	5.9	5.4	7.0	9.1	5.5	6.0	4.5	-0.9	-10.5	-20.4	-18.0	-8.7	-5.8
12 Gross savings of enterprises as percentage of aggregate income payments to individuals	19.6	23.5	19.9	15.8	18.8	18.0	19.3	21.9	18.3	18.8	17.3	12.4	3.5	-5.2	-2.8	6.5	8.8

[25]

covered by data in which it was largest (Table 6, note 7, and Appendix Table III).

By utilizing the estimates prepared by Solomon Fabricant [14] it was possible to correct for the difference between the depreciation and depletion deductions in reproduction and original cost prices for all the years in the period, but only for national income as a whole, not for its various industrial categories.

As shown in Part Four of *Studies in Income and Wealth,* Volume I, the disparity between the value of inventory commodities consumed at cost and at the market price can be estimated roughly by reducing the year-end inventories for each year to a common price level, obtaining the change in inventories, expressing this change in prices for the current year, and then establishing the disparity between this change and the difference in the year-end inventories in their changing current valuations. The adjustment of business savings for this disparity is tantamount to equating the cost of inventory commodities consumed to the sum of: (a) commodity volume of purchases during the year, multiplied by prices paid at the time of purchase, and (b) net change in the commodity volume of year-end inventories, multiplied by the average price prevailing during the year. But the theoretically correct cost of inventory commodity consumed would be equal to the commodity volume under (a), plus the commodity volume under (b), the sum multiplied by prices prevailing at the point of consumption, i.e., when the enterprise succeeds in realizing its share in the national income. It may thus be seen that adjustment for the disparity as described above eliminates from business savings the net profit or loss due to holding of inventories largely during the period elapsing between purchase and utilization, but not during the period of utilization itself; and, because of the use of average prices for the year, yields only approximate results even for that item. Nevertheless, it represents a substantial step towards a theoretically more consistent measurement of the national income produced, and has, therefore, been applied. It could be computed not only for the national product as a whole but also for most of the industrial branches distinguished. Appendix A contains all these adjustments as well as the net savings of business enterprises as they appear before they are adjusted.

[14] His preliminary estimates were published in Measures of Capital Consumption, 1919–1933, *Bulletin 60.* In this report we have utilized his revised estimates which will appear in detail in a final report, *Capital Consumption.*

Finally, savings of government were computed directly, by comparing the net change in the basic fixed assets of government, i.e., buildings, roads, etc., with the net change in total governmental debt. The first item was obtained from estimates of total public construction, available in the capital formation study, reduced for each year by the depreciation on government fixed assets as estimated by Solomon Fabricant.[15] The second item is easily derived from the data on total outstanding governmental debt. The resulting difference represents a crude estimate of net savings by all governmental agencies, i.e., the amount by which the net addition to fixed assets exceeded or fell short of the total change in debt outstanding.

Having commented upon the meaning of the various categories in Table 6, we may now note how gaps in the available data affected the allocations. First, absence of relevant data made it impossible to segregate wages and salaries from entrepreneurial withdrawals in service and miscellaneous industries before 1929. In 1929 total entrepreneurial withdrawals in these two industrial divisions amounted to about 5.1 billion dollars while salaries and wages were 8.5 billion; we decided, therefore, to include the combined item with wages and salaries rather than with entrepreneurial withdrawals. Since total salaries and wages of all industries in 1929 were over 50 billion dollars, the addition of entrepreneurial withdrawals in service and miscellaneous industries could not greatly affect either the relative share or the changes in the relative importance of employees' compensation in the national aggregate of income payments. But the exclusion of this item from total entrepreneurial withdrawals may have affected the latter significantly. Second, as already indicated, net savings of business enterprises prior to 1929 could not be adjusted for gains and losses on the sale of capital assets, an omission that in view of substantial price changes, especially in the early years of the period, may well have affected this item considerably.

2 IMPORTANCE OF SAVINGS OF ENTERPRISES

Gross savings of enterprises represent the share of gross national product retained after all outlays on the costs of production (excluding current consumption of durable capital goods) and all income payments to individuals have been made. The item thus measures that part of the product the final

[15] See below, Section VIII, Table 13.

disposition of which is in the hands of enterprises, and indicates the area in which the decisions of enterprises concerning replacement of capital goods or investment in new goods can be made without the restrictions usually connected with the necessity of obtaining outside funds. The portion of gross national product accounted for by these gross savings of enterprises is fairly substantial, averaging about 16 per cent before 1930, but declining drastically during the depression, and becoming negative in 1932 and 1933 (Table 6, Part B).

Net savings of enterprises are much smaller on the average than gross savings and are much more variable over time. The cumulative total of net savings of business enterprises over the period as a whole, including 1935, amounts to only 4.2 billion dollars; the inclusion of net savings of government raises it to 10.1 billion dollars. These totals average per year only 0.25 and 0.59 billion dollars, respectively, and constitute only 0.4 and 0.9 per cent of the average national income.

However, these averages conceal the extreme variations in net savings of enterprises, and their close conformity to fluctuations in general business conditions—two characteristics that make net savings of enterprises exceedingly important in determining changes in total national income and render its measurement useful in arriving at an understanding of economic fluctuations. The great variability of savings of enterprises is evident when we compare the algebraic average—only 0.59 billion dollars per year (including savings of government), and 0.25 billion dollars when confined to savings of business enterprises alone—with the corresponding averages when signs are disregarded: 4.6 and 3.7 billion dollars per year. The range of variation is quite striking. Total net savings of enterprises ranged from a peak of plus 6.7 billion dollars in 1926 to a trough of minus 10.2 billion dollars in 1932. Of the decline in total national income of 39.8 billion dollars between these two years, over 42 per cent is accounted for by the decline in net savings alone. Similarly, net savings of business enterprises (excluding savings by government) range from a peak of plus 6.7 billion dollars in 1919 to a trough of minus 8.8 billion dollars in 1932; and of the decline in national income of some 20 billion dollars between these two years, the drop in net savings of business enterprises alone accounts for over 15 billion dollars, or 76 per cent. Also the fluctuations in net savings, particularly in savings of business enterprises, follow closely the changes in general business conditions, showing very conspicuously the contractions of 1920–21, 1924, 1927 and 1930–32. It is surprising that both total savings of enterprises and savings of business enterprises declined from 1928 to 1929, but this may be due largely to the adjustment for gains on the sale of capital assets in the latter year but not in the former.

3 CHANGES IN DISTRIBUTION BY TYPE OF PAYMENT

In considering the various types of payment and grouping them in the broad categories of employees' compensation, entrepreneurial income payments and property income, we are concerned not only with the absolute quantities but also with the relative proportions accounted for by each. Changes in these proportions are especially important because these types of income represent largely the compensation of various groups in the economic system. Wages and the bulk of salaries constitute the main income of a large group in the population whose per capita income is relatively low; and changes in the relative share of wages and salaries indicate, though only approximately, changes in the relative share of the low income groups. Dividend and interest disbursements are received largely by those whose average income is relatively high; and a change in the proportion of income going out as dividends and interest is, with certain qualifying conditions, an index of the change in the share of income received by the high income groups. Entrepreneurial income payments occupy an intermediate position, and tend to vary greatly in average magnitude from one industrial branch to another. Dominated by agriculture and retail trade, they represent the main income of that large class of small entrepreneurs whose average income is fairly low and is subject to the vicissitudes of the competitive struggle.

The share of various types of payment is measured in percentages of aggregate income payments, not in percentages of national product, for savings of enterprises cannot be assigned as an income share to any specific group of income recipients. There is often an inclination to treat net savings as part of property income, on the ground that such savings or losses affect most conspicuously the value of securities held by the recipients of interest and dividend payments. But if incurring a net business loss or retaining net business savings by an enterprise does affect the fortunes of its security holders, the net savings as measured by us are hardly a measure of the effect. And it may reasonably be argued that savings of enterprises in an industry are as important to the employees as to the owners, and have as much effect on their economic welfare. Upon this assumption, percentage allocation among the types of payment of national income would be identical with that of aggregate income payments to individuals.

Employees' compensation, including wages and salaries, pensions, relief and entrepreneurial withdrawals in service and miscellaneous industries, accounted on the average for about 70 per cent of the total flow of income payments to individuals; and this relative share remained fairly constant over the period (Table 6, Part B, and Chart 4). If the percentages are averaged for the two halves of the post-War period, the resulting means suggest a slight upward trend. This impression is confirmed when it is observed that before 1926 the percentage in the third line never went above 69.8, while beginning with 1926 it was at the level of 69.9 or higher in all years except one. But the differences are too small to indicate any definite rise in the share of employees' compensation; and the averages are considerably affected by the increased importance of relief in 1933, 1934 and 1935.

While no significant movement over the period as a whole can be observed in the relative share of employees' compensation, the other two major groups of payment both show distinct and significant trends. The share of entrepreneurial income payments declined continuously, owing largely to the decline in the relative share of entrepreneurial withdrawals. Indeed, prior to 1932, the percentage share of the latter declined each year except in 1924, and from a level of about one-sixth of total income payments in 1919 and 1920 it dropped to about one-tenth in 1933 and 1934.[16] The relative share of the rent item showed at first a rise, associated with the decline in expenses and the high rent levels prevailing in the years before expansion in the volume of construction exercised a moderating effect on rents; then a decline, which became especially precipitous after 1929. In contrast to entrepreneurial income payments, the relative share

Chart 4
VARIOUS TYPES OF PAYMENT AND SAVINGS OF ENTERPRISES IN PERCENTAGES OF AGGREGATE INCOME PAYMENTS
1919 – 1935

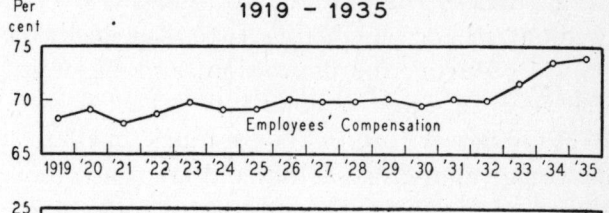
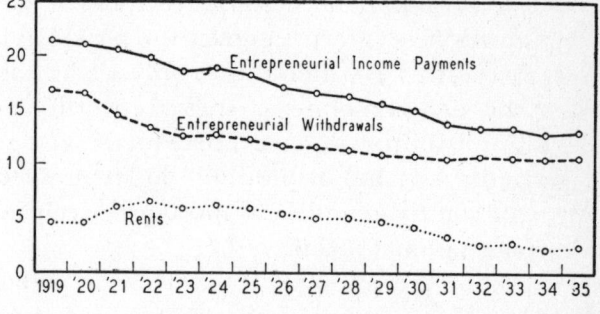
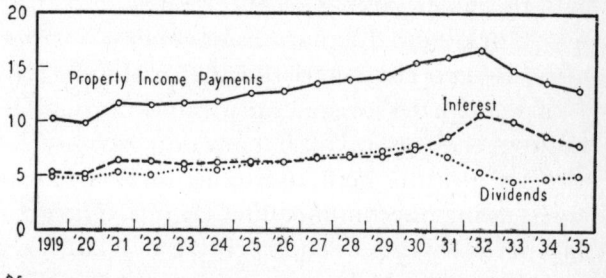
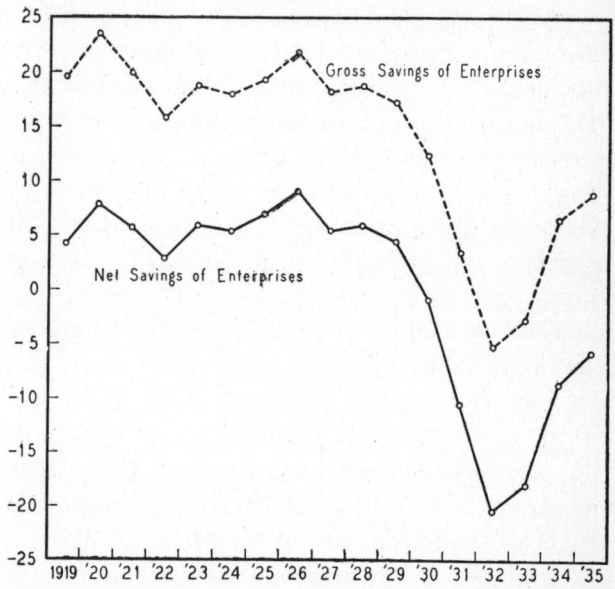

of property income rose distinctly over the period as a whole, accounting for one-tenth of total income payments in the early years of the period and about one-sixth in the later years. This upward tendency was manifest in both dividend and interest payments before 1929, but since then primarily

[16] This downward tendency would undoubtedly persist even were entrepreneurial withdrawals in service and miscellaneous industries included with all other entrepreneurial withdrawals. If we assume that this item, now included with employees' compensation, doubled between 1919–20 and 1929, an assumption which tends to exaggerate any upward movement that may have existed in it, the inclusion would add slightly over 4 per cent in 1919–20 and slightly over 6 per cent in 1929 to the share of entrepreneurial income payments. This would still leave a decline in entrepreneurial withdrawals from about 21 per cent in 1919–20 to 17 in 1929.

On the other hand, the exclusion of this item from employees' compensation would remove completely any slight rise in its relative share; this, of course, on the improbable assumption that the increase in entrepreneurial withdrawals in service and miscellaneous industries attained the striking magnitude suggested above.

in the latter. If rents were added to dividend and interest payments to obtain a new total of property income, the share of the latter would still show a marked rise over the period.

The changes in the relative shares of various types of payment indicated in Table 6 are of considerable significance, and deserve further exploration, especially to ascertain whether both the changes over the period and the fluctuations that appear to be associated with business cycles persist when the dollar volumes are corrected for changes in the price level. But an adjustment for changes in the *general* price level would leave the percentage distribution of Table 6 undisturbed. To make the test significant, we need measures that would distinguish the prices of commodities and services purchased by salary and wage earners from the prices of commodities and services bought by recipients of entrepreneurial and property income payments. The absence of such price data postpones this inquiry.

The conclusions from Table 6 can be further elaborated by studying the distribution within the various industrial branches. We turn now to the cross-classification by type of payment and by industrial source.

V CROSS-CLASSIFICATION BY TYPE OF PAYMENT AND BY INDUSTRIAL SOURCE

IN CONSIDERING the combined classification by type of payment and by industrial source, we might pursue two lines of inquiry. We might determine whether the changes observed in the relative shares of various types of payment in the national total are also present within each industrial branch distinguished; and ascertain how the shift in the relative importance of various industrial branches contributed to the changes in the relative proportions of the various types of payment. We might study the changes observable in the relative share of various industrial branches within each type of payment; and see how the shift in the relative importance of various types of payment contributed to the changes in the relative share of the various industrial branches.

It would not be possible here, or generally profitable, to pursue both lines of inquiry. We confine the analysis to the effect of the change in the distribution by type of payment within each industrial branch, and of the shift in the relative share of the various industrial branches, on changes in the percentage share of the various types of payment in aggregate income payments. We thus make the changes in the distribution by type of payment the dependent variable, rather than the shifts in the relative share of various industrial divisions in the economic system: the growth and change of the various industries appear to us to be the independent factor that provides the framework within which changes in the percentage distribution by type of payment are to be understood.

The first question to be answered is whether shifts in the relative shares of various types of payment in the total also occur in each industrial branch. Do the slight upward movement in the share of employees' compensation, the downward movement in the share of entrepreneurial income payments, and the marked upward movement in the share of property income appear in the distribution of income payments originating in each industrial division? An answer to this question is provided in Table 7, which measures the movement over the period in the relative share of the various types of payment by the total change in the average percentage for two pairs of segments of the post-War period: 1919–26 and 1927–34, and 1922–26 and 1927–31. We have computed similar measures also for the two intervening sets of periods (omitted in the table); our conclusions are based on the consideration of all these measures.[17]

The relative share of employees' compensation, which for the country as a whole rose slightly, appears to have declined significantly in a majority of the industrial categories. Indeed, if we include the subdivisions of public utility and finance, it declined in nine of the fifteen industrial branches distinguished, and rose in only three (trade, government and real estate).

[17] For a more extended table of this type, confined to nine industrial branches, see *Bulletin 59*, Table 6, p. 15; see also Table 8 below.

Table 7

CHANGE IN PERCENTAGE SHARE OF VARIOUS TYPES OF PAYMENT WITHIN EACH INDUSTRIAL BRANCH

Industrial branches	Average volume of all income payments (millions of dollars) 1919-34	Compensation of Employees			Entrepreneurial Income Payments			Property Income Payments		
		Average percentage of all income payments for industry 1919-34	Total change in average percentage share from		Average percentage of all income payments for industry 1919-34	Total change in average percentage share from		Average percentage of all income payments for industry 1919-34	Total change in average percentage share from	
			1919-26 to 1927-34	1922-26 to 1927-31		1919-26 to 1927-34	1922-26 to 1927-31		1919-26 to 1927-34	1922-26 to 1927-31
(1)	(2)	(3)	(4)	(5)	(6)	(7)	(8)	(9)	(10)	(11)
1 Agriculture	6,609	16.3	-1.8	0.0	77.7	+0.4	-0.1	6.0	+1.3	+0.1
2 Mining	1,780	84.1	-4.0	-3.4	1.4	-0.1	-0.2	14.5	+4.2	+3.6
3 Manufacturing	13,960	83.9	-2.9	-3.0	2.3	-0.7	-0.5	13.7	+3.6	+3.6
4 Construction	2,133	80.6	-2.0	+2.9	17.1	+0.8	-3.8	2.3	+1.2	+0.9
5 Transportation and other public utilities										
a Electric light and power and manufactured gas	6,535	72.9	-10.5	-6.2	0.1	0.0	0.0	27.0	+10.6	+6.3
b Steam railroads, Pullman and express	981	42.5	-14.0	-8.9	0.3	-0.4	-0.2	57.2	+14.4	+9.1
c Pipe lines, street railways and water transportation	3,712	78.7	-5.6	-3.7	0.0	0.0	0.0	21.3	+5.6	+3.7
d Communication	1,154	80.2	-3.1	-1.4	0.4	-0.1	+0.1	19.4	+3.2	+1.3
	688	77.2	-6.5	-2.2	0.0	0.0	0.0	22.8	+6.5	+2.2
6 Trade	8,728	71.2	+1.9	+0.9	24.1	-2.1	-1.7	4.8	0.0	+0.7
7 Finance	7,157	31.6	+7.3	+5.1	42.3	-18.7	-14.1	26.1	+11.4	+9.0
a Banking	871	63.4	+2.8	-1.5	0.0	0.0	0.0	36.6	-2.8	+1.5
b Insurance	924	104.3	-1.4	-2.0	0.0	0.0	0.0	-4.3	-1.4	+2.0
c Real estate	5,362	13.7	+2.9	+3.3	56.1	-20.8	-15.7	30.2	+18.0	+12.5
8 Government	6,525	77.5	+3.5	+3.5	0.0	0.0	0.0	22.5	-3.5	-3.5
9 Service	8,259	98.4	-0.5	-0.5	1/	—	—	1.6	+0.5	+0.5
10 Miscellaneous	2,588	90.7	+0.4	-0.8	1/	—	—	9.3	-0.4	-0.8
11 Total	64,274	69.8	+1.6	+0.5	17.0	-4.8	-3.0	13.2	+3.3	+2.5

1 Included with employees' compensation.

The share of entrepreneurial income payments declined distinctly, both as a percentage of aggregate income payments and within the various industrial divisions (Table 7). If we include the subdivisions there are eight industrial branches in which unincorporated enterprises are significant and in which entrepreneurial income payments can be segregated. Of these eight, there is a decline in the relative share of entrepreneurial income payments in five, and a distinct rise in none. However, in agriculture, the industry that constitutes the most important source of entrepreneurial withdrawals, their share in total income payments showed no definite decline; indeed, in three of the four sets of periods compared it rose. Obviously, for entrepreneurial income payments as for employees' compensation, the movement of the share has been considerably affected by the shift in the relative position of the various industrial branches.

The percentage share of property income rose in twelve of the fifteen industrial branches distinguished in Table 7. In two, banking and miscellaneous, the general direction of the movement is not quite clear; in only one, government, does it decline definitely over the period. This evidence of Table 7 does not necessarily mean that the shift in the relative importance of various industries has not contributed to the rising tendency observed in the share of property income. But it does indicate that this upward movement would have appeared even had the relative share of each industrial branch remained constant.

The analysis may be developed to the point of actual measurement of the extent to which the movements in the relative shares of various types of payment have been affected, on the one hand, by changes within the industrial branches (intra-industry shifts) and, on the other hand, by changes in the relative weight of the various industries (inter-industry shifts). The effect of intra-industry shifts can be isolated by deriving for each type of payment a series of percentages representing what its relative share in total income payments would have been in each year had the relative importance of each industrial branch remained constant over the period. This series can be computed by obtaining for each type of payment for each year a weighted average of its relative share in each industrial branch, the weights being the *average* volume for the period of the income payments originating in each industrial branch.

Similarly, the effect of inter-industry shifts can be isolated by deriving for each type of payment a series of percentages representing what its relative share in total income payments would have been in each year had the percentage distribution among the various types of payment remained constant for each industrial branch over the period. This series can be computed by obtaining weighted averages of the same type as those described above, except that here the percentages averaged are kept constant at their average value for the period, while the weights vary, representing, for each year, the actual volume of income payments originating in each industrial branch.

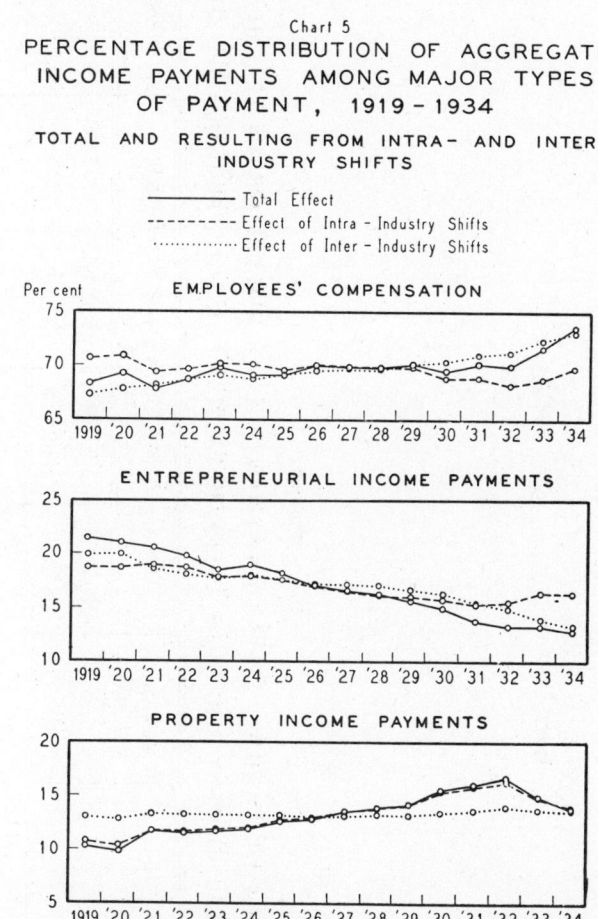

Chart 5
PERCENTAGE DISTRIBUTION OF AGGREGATE INCOME PAYMENTS AMONG MAJOR TYPES OF PAYMENT, 1919-1934
TOTAL AND RESULTING FROM INTRA- AND INTER-INDUSTRY SHIFTS

Such series have been computed and are presented graphically in Chart 5. For each we obtained measures of the movement over the period as a whole. These measures, together with those obtained for the percentage shown in Table 6, are brought together in Table 8. The measures are given for all the sets of periods compared, and include not only the total change, but also the same changes reduced to a per year basis.

Table 8 shows clearly the effects of the two sets of factors on the movements over the period in the relative share of various types of payment. The slight rise in the relative share of employees' compensation in the national total is due exclusively to the shift among various industries, i.e., to a net relative gain in the weight of these industrial di-

Table 8

CHANGE IN PERCENTAGE SHARE OF VARIOUS TYPES OF PAYMENT IN AGGREGATE INCOME PAYMENTS TO INDIVIDUALS, TOTAL AND RESULTING FROM INTRA-INDUSTRY SHIFTS IN TYPES OF PAYMENT AND INTER-INDUSTRY SHIFTS IN TOTAL PAYMENTS

Type of payment	Average percent-age share 1919-34	Total change in average percentage share from				Change per year in average percentage share from			
		1919-26 to 1927-34	1920-26 to 1927-33	1921-26 to 1927-32	1922-26 to 1927-31	1919-26 to 1927-34	1920-26 to 1927-33	1921-26 to 1927-32	1922-26 to 1927-31
Compensation of Employees									
Total (inter- and intra-industry change combined)	69.8	+1.5	+1.1	+0.9	+0.5	+0.2	+0.2	+0.2	+0.1
Due to intra-industry shifts only	69.7	-0.8	-0.8	-0.6	-0.5	-0.1	-0.1	-0.1	-0.1
Due to inter-industry shifts only	69.8	+2.4	+1.9	+1.5	+1.1	+0.3	+0.3	+0.2	+0.2
Entrepreneurial Income Payments									
Total (inter- and intra-industry change combined)	17.0	-4.8	-4.3	-3.7	-3.0	-0.6	-0.6	-0.6	-0.6
Due to intra-industry shifts only	17.1	-2.1	-2.1	-2.1	-1.8	-0.3	-0.3	-0.4	-0.4
Due to inter-industry shifts only	17.0	-2.7	-2.2	-1.6	-1.1	-0.3	-0.3	-0.3	-0.2
Property Income Payments									
Total (inter- and intra-industry change combined)	13.2	+3.3	+3.3	+3.0	+2.5	+0.4	+0.5	+0.5	+0.5
Due to intra-industry shifts only	13.2	+2.9	+2.9	+2.6	+2.2	+0.4	+0.4	+0.4	+0.4
Due to inter-industry shifts only	13.3	+0.4	+0.3	+0.2	+0.2	0.0	0.0	0.0	0.0

visions in which the share of employees' compensation is high. The intra-industrial shift in the relative share of employees' compensation is, on the contrary, distinctly downward. In the general contraction of the relative share of entrepreneurial income payments the inter and intra-industrial shifts both exercised effects in the same direction and were almost equally important. The rise in the relative share of property income is also due both to its rise within the various industries and to the shift among industries; but the inter-industrial shifts have in this case exercised less effect.

These conclusions, especially that concerning the movement of the share of employees' compensation, shed light on the changes during the post-War years in the functional and industrial structure of income in this country. The considerable shift in the distribution of employment and service income among groups attached to various industrial branches was concealed by its stability when expressed as a percentage of the national total. Obviously such a shift in the composition of employees' compensation, if accompanied by immobility of employment and income among the various industrial groups, might lead to serious problems of social adjustment.

The potential significance of this shift is revealed clearly in Table 9, in which the average percentage distribution of total employees' compensation among the various industrial branches is compared for the first and second halves of the post-War period. For comparative purposes, similar averages are provided for the industrial distribution of total income payments to individuals.

It is clear from Table 9 that, appreciable as was the shift in total income payments from the commodity producing to the service industries, the same shift in total employees' compensation was still more marked. The commodity producing industries and transportation plus other public utilities accounted during the first half of the period for 51 per cent of total employees' compensation; in the second half of the period for only 41 per cent. And of course during the recent depression this change became still more striking. If we assume that the wage levels in the two groups of industries move similarly, this change in the distribution of the wage bill means a shift from wage earning labor in basic industries to white collar, salaried labor and to such secondary labor supply as provides most of the man-power in trade and some of the service industries.

Also, in such industries as mining, manufacturing and steam railroads, in which a clear distinction between wages and salaries can be made, the share of wages declined over the period, while the share of salaries rose, with corresponding differences in the movement of per capita wages and salaries.[18] It thus appears that over the period as a whole the income of wage earners in the basic industries constituted a declining share of national income.

[18] See *Bulletin 59*, Table 7, p. 18.

Table 9

EMPLOYEES' COMPENSATION AND AGGREGATE INCOME PAYMENTS, PERCENTAGE DISTRIBUTION AMONG INDUSTRIAL BRANCHES

Averages for the Period and for the Two Halves of the Period

INDUSTRIAL BRANCHES	COMPENSATION OF EMPLOYEES				AGGREGATE INCOME PAYMENTS			
	AVERAGE 1919-34	AVERAGE 1919-26	AVERAGE 1927-34	CHANGE FROM 1919-26 TO 1927-34	AVERAGE 1919-34	AVERAGE 1919-26	AVERAGE 1927-34	CHANGE FROM 1919-26 TO 1927-34
1 Agriculture	2.4	2.9	1.9	−1.0	10.3	11.7	8.8	−2.9
2 Mining	3.3	4.1	2.5	−1.6	2.7	3.3	2.2	−1.1
3 Manufacturing	26.0	28.1	23.8	−4.3	21.5	22.7	20.4	−2.3
4 Construction	3.8	4.2	3.3	−0.9	3.2	3.5	2.9	−0.6
5 Transportation and other public utilities	10.7	11.7	9.6	−2.1	10.2	10.3	10.0	−0.3
6 Trade	13.8	13.9	13.8	−0.1	13.6	13.7	13.5	−0.2
7 Finance	5.0	4.4	5.5	+1.1	11.0	10.9	11.2	+0.3
8 Government	11.6	9.8	13.5	+3.7	10.5	8.9	12.0	+3.1
9 Service	18.2	16.2	20.1	+3.9	12.9	11.4	14.5	+3.1
10 Miscellaneous	5.2	4.7	5.8	+1.1	4.0	3.6	4.5	+0.9
11 Total	100.0	100.0	100.0		100.0	100.0	100.0	

VI THE DISTINCTION BETWEEN CONSUMERS' OUTLAY AND CAPITAL FORMATION

1 THE NATIONAL PRODUCT, CONSUMERS' OUTLAY AND CAPITAL FORMATION

IN SECTION I we noted that the two comprehensive totals subsumed under the term 'national product' could be defined at the production, distribution or consumption phases of economic circulation. The apportionment of the national product at the production phase, that according to industrial origin, and at the distribution phase, that according to type of income and income payment, have already been discussed. We now consider its apportionment at the stage of consumption. But first we must redefine national product in order to draw distinctions made possible by the available data. The third set of definitions in Section I, which equated national product to the value of goods consumed by individuals plus savings of individuals and enterprises, is not satisfactory for the present purpose because it suggests a distinction drawn at the point of actual consumption of goods by individuals, a distinction which the lack of data prevents us from following; and because it analyzes national product partly on the goods level and partly on the flow-of-funds level.

If only the goods content is considered, economic activity appears as a continuous flow of commodities and services that change their form and enter one another in a successive alternation of consumption and production. The reference of the concepts of net and gross national product to this continuous flow implies the introduction of some fixed point, at which the measurement is made and the net and gross value of goods gauged. Commodities and services that reach this point may be considered finished and at their destination, and it is their value that largely determines the magnitude of the national product.

As suggested in the third set of definitions of national product in Section I, this point is fixed at the level at which commodities and services are in the form and at the place that render them immediately available for use by the individuals who comprise the nation. It is this use by the final consumers, i.e., the households [19] of the nation that provides the fixed point of reference at which the continuous flow of economic goods may be viewed and their volume measured. True, the national product for any given time unit is not fully accounted for by the flow of finished goods to their destination, i.e., by commodities and services ready for use by final consumers and reaching their hands. Some portion of the nation's economic activity may result in an increase in the stock of finished commodities in hands other than those of final consumers and in the stock of unfinished commodities; or, if we take into consideration the international aspects of the economy, in a movement of commodities and services abroad. But these are the only other elements in the national product, which may thus be redefined as equal to: (a) the value of finished commodities and services reaching their destination; (b) changes in stocks of all finished commodities not at their destination (in circulation) and of all other commodities; (c) the net change in claims against foreign countries, this item representing the fullest measure of the net product flowing out of the country's boundaries.

The difference between net and gross national product in terms of this definition lies in the scope of item (b), changes in stocks of commodities. When net national product or national income is considered, this item covers *net* changes in stocks of commodities in the hands of enterprises, whether they are raw materials, semifinished products, finished commodities, or fixed capital equipment. When gross national product is considered, the item covers net changes in stocks of raw materials, semifinished products and finished products in the hands of enterprises; but in the case of fixed capital equipment *gross* rather than *net* changes in stocks are taken, i.e., we include the gross addition to the volume of such equipment in existence at the beginning of the time unit, without any allowance for its current consumption. And again, as mentioned above, fixed capital equipment, in both concepts, includes residential real estate whether owned by corporations or individuals.

The distinction between consumers' outlay and capital formation follows this new set of definitions naturally. In the case of net national product or national income, item (a) represents consum-

---

[19] This term as used here and subsequently also includes larger groups of consumers, i.e., hospitals, homes for aged, dormitories, hotels, restaurants.

ers' outlay, and the sum of items (b) and (c) represents net capital formation. For gross national product, item (a), which is identical in scope with item (a) for national income, represents consumers' outlay; and the sum of item (b), now different, and item (c), the same in both concepts, represents gross capital formation. The difference between net and gross national product is thus made to persist in the difference between net and gross capital formation, the concept and volume of consumers' outlay remaining the same in both national product totals.

Two aspects of the distinction just drawn between consumers' outlay and capital formation deserve notice. First, the distinction is significantly different from that between consumption and additions to the stock of wealth. In the latter, consumption represents actual consumption by ultimate consumers, and not as in the case of consumers' outlay, their acquisition of finished products. Correspondingly, additions to stock is a broader category than capital formation, since it includes, besides the contents of the former, changes in stocks of finished commodities in the hands of ultimate consumers. Interesting as the apportionment between consumption and additions to the stock of wealth would have been, there are no data available to measure it on a continuous and satisfactory basis.

Second, in the distinction between consumers' outlay and capital formation, item (b) is restricted to changes in stocks of *commodities*, and thus excludes any possible changes in stocks of intangible property. Obviously, enterprises may consume commodities and services in order to enhance good will, the preference demand for their product, or some valuable rights, and not all of this consumption may be reflected in the value of finished commodities flowing to ultimate consumers. Moreover, individuals may purchase currently produced services not for consumption in the general process of living but in order to improve themselves as economic producers. Both cases represent a diversion of the current product of the economy from flow to ultimate consumers as ultimate consumers, and should properly be included under item (b) of the definition, and thus enter capital formation. But the data available for the measurement of these items in capital formation are lacking, their segregation is theoretically difficult, and they are thus necessarily included in consumers' outlay when its magnitude is obtained by subtracting capital formation, directly measured, from the national product. It may be said, however, that the items are relatively small; and that in the second of the two the demarcation between purchases of ultimate consumers as consumers and their purchases for the purpose of increasing their personal capacity as producers is inconsistent with the concept of national product used in this report.

The statistical study, whose results make possible the apportionment of the national product between consumers' outlay and capital formation, aimed at a direct measurement of capital formation; the volume of consumers' outlay given in the tables below has been obtained by subtraction. Since the measurement of capital formation required a more explicit definition of the concept than that given above, a more detailed discussion of the measures of capital formation is needed before the estimates themselves and the apportionment between consumers' outlay and capital formation can be clearly understood.

2 METHODS OF ESTIMATION AND THE CLASSIFICATION OF COMMODITIES

In estimating the annual volume of gross and net capital formation as defined above, two methods are possible. One, the change-in-stock method, proceeds to establish at some time within each year the stock of commodities as defined under (b) for net national product. Then by a process of subtraction, the net change in the stock can be obtained, and the addition of the net change in claims against foreign countries yields directly the volume of net capital formation. If to this volume is added the estimated current consumption of fixed capital equipment (inclusive of residential real estate) the total represents gross capital formation.

The other method combines the flow-of-goods method with the change-in-stock method, the former accounting for quantitatively the most important part of the final estimates. It begins with the measurement of the flow of commodities, rather than of the stock in existence. In the current flow of various commodities and services it attempts to segregate those which, because of their technical nature, are bound to constitute a gross addition to the stock of fixed capital equipment in the hands of their users. Provided such a segregation can be made, the resulting total will account for the bulk of gross capital formation. The items missing would be the net change in claims against foreign countries and the net change in stocks of finished and unfinished commodities, exclusive of fixed capital equipment at its destination. With these two additional items provided, the sum will cover gross capital formation fully. From this sum is subtracted the current consumption of fixed capital equipment to obtain net capital formation.

This combination of the two methods, actually followed in the study of capital formation, demands first of all a careful classification of commodities to make possible the segregation of the net output without duplication, except that involved in the consumption of fixed durable equipment; and then the segregation, within the net output, of those commodities that must be a part of capital formation. The commodity classification used is given in detail below. It is designed to

CROSS-CLASSIFICATION OF COMMODITIES
(for definition of terms used see footnote 20)

A Consumers' Goods

I *Perishable*
1 *At Destination*
  a Finished—bread, coal used by households, etc., in hands of households
  b Unfinished—none

2 *In Circulation*
  a Finished—same as under AI–1a, but in hands of producers and distributors
  b Unfinished—raw materials, fuels, supplies so far as they are used for production, transportation and distribution resulting in AI–1a

II *Semidurable*
1 *At Destination*
  a Finished—shoes, clothing, etc., in hands of households
  b Unfinished—none

2 *In Circulation*
  a Finished—same as under AII–1a, but in hands of producers and distributors
  b Unfinished—raw materials, fuels, supplies, used for production, transportation and distribution resulting in AII–1a

III *Durable*
1 *At Destination*
  a All finished—passenger cars, jewelry, furniture in hands of households
    aa Residential buildings
    bb All other
  b Unfinished—none

2 *In Circulation*
  a Finished—same as under AIII–1a, but in hands of producers and distributors
  b Unfinished—raw materials, fuels, supplies, used for production, transportation and distribution resulting in AIII–1a

B Producers' Goods

I *Perishable*
None

II *Semidurable*
None

III *Durable*
1 *At Destination*
  a Finished—industrial, farm machinery, buildings, trucks, etc., in hands of business units who will use them
  b Unfinished—none

2 *In Circulation*
  a Finished—same as under BIII–1a, but in hands of producers and distributors
  b Unfinished—raw materials, fuels, supplies, etc., used for production, transportation and distribution resulting in BIII–1a

[20] *Durable Commodities*—Commodities that, without marked change and retaining their essential physical identity, are ordinarily employed in their ultimate use over a long period (for purposes of this study more than three years). Examples: a building; a steam engine or dynamo; an automobile or truck; a bed, table or chair.
Non-durable Commodities—Commodities that, without marked change and retaining their essential physical identity, are ordinarily employed in their ultimate use over a short period (for purposes of this study less than three years). Non-durable commodities are further classified into:
Semidurable Commodities—Commodities that, without marked change and retaining their essential physical identity, are ordinarily employed in their ultimate use for from six months to three years. Examples: automobile tires, clothing, shoes.
Perishable Commodities—Commodities that, without marked change and retaining their essential physical identity, are ordinarily employed in their ultimate use less than six months. Examples: bread, cigarettes.
Commodities at Destination—Commodities that have reached either the household or the producing units wherein they find their ultimate use. Examples: bread in the household larder; truck in hands of firm using it.
Commodities in Circulation—Commodities that are still in process of production, transportation or distribution and have not as yet reached the units wherein they find their ultimate use. Examples: clothing in department store; coal in factory bin.
Finished Commodities—Commodities, whether durable or non-durable, in the form in which, without significant alteration, they are employed in their ultimate use. Examples: shoes, furniture, machinery.
Unfinished Commodities—Commodities that, whether ultimately durable or non-durable, are not yet in the form in which they are employed in their ultimate use. Examples: raw cotton; structural steel beams.

facilitate the measurement of gross capital formation but it also includes some classification of commodities that enter consumers' outlay.

The classification refers to commodities alone, not to services, or rather only to services that have been embodied in new commodities. But one group of services is of some importance in measuring capital formation, viz., services performed in connection with existing finished durable commodities (usually at destination) that neither result in the production of a new finished commodity nor constitute non-durable current maintenance. Such services are typified by a substantial alteration of an existing building; or a substantial repair or reconstruction of an engine already produced and installed. They may be interpreted as constituting additions to stocks of commodities rather than as representing consumers' outlay, and should therefore be considered in the more detailed definition of the scope of capital formation.

3 VARIANTS OF THE DISTINCTION BETWEEN CONSUMERS' OUTLAY AND CAPITAL FORMATION

In the light of the preceding discussion and with the help of the commodity classification we can now outline more comprehensively the composition of consumers' outlay and of capital formation, and describe several possible variants of this distinction. Beginning with the variant that conforms most closely to the apportionment defined in the discussion, we proceed to some modifications of this apportionment.

VARIANT I
(corresponding to definitions given above and describing most closely the apportionment actually measured)

GROSS

GCF I–Gross Capital Formation=
1. flow of finished producers' durable commodities and residential buildings to final domestic users (BIII–1a plus AIII–1aa)
  and
2. net changes in stocks of all commodities in the classification, except AI–1a, AII–1a, AIII–1a, and BIII–1a
  and
3. net change in claims against foreign countries

CO I–Consumers' Outlay=
GNP I (gross national product as defined in this report) minus GCF I=
flow to ultimate domestic consumers of all finished commodities except residential buildings (AI–1 plus AII–1 plus AIII–1bb)
  and
the value of all services not embodied in new commodities

NET

NCF I–Net Capital Formation=
GCF I minus current consumption of producers' finished durable commodities and residential buildings (i.e., of BIII–1a and of AIII–1aa)

CO I–Consumers' Outlay=
NNP I (net national product as defined in this report) minus NCF I=
GNP I minus GCF I

It may be suggested that consumers' durable finished commodities, other than residential buildings, should be treated in the same fashion as the latter, i.e., as fixed capital equipment yielding income. This would expand the concept of gross capital formation to include the flow of the whole group of consumers' durable commodities, but would demand as well a broader concept of national product than that used in this report. This broader concept would include not only income from residential buildings but also from the use of other consumers' durable commodities at their destination. The distinction between consumers' outlay and capital formation that results from this change in concept is Variant II.

Consumers' Goods—Commodities and services that, whether finished or unfinished, are, when finished and at their destination, used by households or large ultimate consuming units. Examples: flour, bread, raw wool, clothing.
Producers' Goods—Commodities and services, whether finished or unfinished, that are, when finished and at their destination, used by business agencies in the process of production. Examples: industrial machinery; steel used therein.

NATIONAL INCOME AND CAPITAL FORMATION

VARIANT II

GROSS

GCF II–Gross Capital Formation=
GCF I plus flow of consumers' durable commodities other than residential buildings (AIII–1bb) to ultimate domestic consumers

CO II–Consumers' Outlay=
GNP II minus GCF II, where GNP II, the new concept of gross national product, is equal to GNP I plus net income originating in the use of AIII–1bb plus current consumption of AIII–1bb. Hence CO II=CO I plus (net income originating in the use of AIII–1bb plus current consumption of AIII –1bb minus flow of AIII–1bb to their ultimate users). If AIII–1bb is purchased only for replacement of current consumption, CO II is larger than CO I by the net income accruing from use of AIII–1bb

NET

NCF II–Net Capital Formation=
GCF II minus current consumption of all producers' and consumers' finished durable commodities at destination=
NCF I plus flow of AIII–1bb minus current consumption of AIII–1bb

CO II–Consumers' Outlay=
NNP II minus NCF II, where NNP II, the new concept of net national product or national income, is equal to NNP I plus net income originating in the use of AIII–1bb

Repairs and alterations of already existing durable finished commodities, whether they are producers' or consumers', are important for the measurement of capital formation only if we assume that the results are durable and hence would not be consumed within the same time unit as they are produced, or immediately following. If these activities are included, it may be assumed that the concept of both capital formation and national product as defined in Variants I and II already allow for the income resulting from these repairs and alterations, past and present. Considering these repairs and alterations in their application to Variant II, we obtain a new and the most comprehensive variant.

VARIANT III

GROSS

GCF III–Gross Capital Formation=
GCF II plus value of repairs and alterations of existing finished durable commodities

CO III–Consumers' Outlay=
GNP II minus GCF III, on the assumption that GNP II already includes net income originating from results of repairs and alterations, past and present, and current consumption of results of repairs and alterations, past and present

NET

NCF III–Net Capital Formation=
GCF III minus current consumption of all durable commodities, including the part assumed to be provided for by repairs and alterations

CO III–Consumers' Outlay=
NNP II minus NCF III, on the assumption that NNP II already includes net income originating from results of repairs and alterations, past and present

This schematic presentation of the three variants indicates briefly the composition of consumers' outlay and of capital formation under various assumptions. The values of all the items are taken at cost to those particular groups which, for the purposes of definition and measurement, are the ultimate holders and recipients of the commodities. Thus the flow of finished commodities and services to final consumers is evaluated at the cost to them. The net changes in stocks of commodities are measured at their cost to those who hold them.

Our study has yielded estimates of gross capital formation corresponding to the first two variants, and an approximation to gross capital formation corresponding to the third variant. But the data on the current consumption of finished durable commodities or on results of repairs and alterations, provided by Solomon Fabricant in his study of capital consumption, make possible an evaluation of net capital formation only in accordance with Variant I. Similarly, from the national income estimates we can measure net and gross national product only in accordance with the concepts used in this report, i.e., corresponding to Variant I. For this reason neither net capital formation as described in Variants II and III, nor consumers' outlay corresponding to these variants, can be measured. In Sections VII–IX we discuss primarily the distinction between consumers' outlay and capital formation and the analysis of these two parts of the national income product, all as described in Variant I. But in Section VII estimates of gross capital formation in accordance with Variant II are also presented, and in Appendix C some data making possible an approximation to gross capital formation in accordance with Variant III are provided.

VII APPORTIONMENT OF GROSS NATIONAL PRODUCT BETWEEN GROSS CAPITAL FORMATION AND CONSUMERS' OUTLAY

1 GROSS CAPITAL FORMATION–CHARACTERISTICS OF THE ESTIMATES

The totals of gross capital formation in the United States for 1919–35, presented in Table 10 in two variants, correspond in coverage to the composition of gross capital formation given in the outline of Variant I and Variant II. There were, however, some reinterpretations of the theoretical definition, and some departures from it, the latter caused by the exigencies of the available data.

As defined above, gross capital formation in either Variant I or II excludes all repairs and alterations. But in one industry, viz., construction, there is a considerable volume of alteration and rebuilding activity whose result is tantamount to new construction. It seemed illogical to include new construction and to exclude these substantial alteration jobs which with respect both to the cost and durability of results are little different from new construction. Accordingly, such substantial alterations of construction units as called for building permits were treated as new construction and are included in the estimates of the volume of construction appearing in Table 10 and subsequently. Other repairs and alterations were excluded, but their volume, so far as data were available, was estimated (see Appendix C, Table VII).

Parts sold for replacement in finished durable commodities at their destination raised a similar problem. Here again, especially in the case of machinery and equipment used in the process of production, parts may be conceived as having a life as long as or not much shorter than the commodity when new; and the total of many machines may be treated as a combination of parts since they are ordinarily replaced part by part until the framework is the only original piece that survives. It therefore appeared advisable to expand the concept of producers' finished durable commodities to include parts sold for replacement. The case seemed much weaker for parts of consumers' durable commodities; they have consequently been treated as unfinished.

We may now consider how the availability or lack of data affected the estimates given in Table 10. The flow of movable commodities, producers' and consumers' taken together, is reasonably complete, covering all manufactured commodities (as reported in the Census of Manufactures) and whatever finished durable commodities flow directly from agriculture and mining. The item omits the insignificant volume of durable commodities produced in the service industries. The volume of construction, inclusive of such alterations and repairs as call for building permits, is based upon a comprehensive estimate prepared for nonfarm res-

Table 10

GROSS CAPITAL FORMATION, BY TYPE OF USER, 1919-1935

(millions of dollars)

	1919	1920	1921	1922	1923	1924	1925	1926	1927	1928	1929	1930	1931	1932	1933	1934	1935
							Current Prices										
Destined for use by																	
1 Consumers																	
a Residential construction	1,732	1,493	2,241	3,524	4,422	4,713	5,202	4,757	4,524	4,255	3,010	1,805	1,262	444	392	458	923
2 Business	13,128	16,681	6,166	7,165	11,583	7,558	11,137	11,668	10,402	9,916	13,903	8,152	4,393	655	1,858	2,680	5,095
a Flow of producers' durable commodities	6,234	6,177	3,926	3,848	5,267	4,962	5,287	5,716	5,461	5,852	6,908	5,480	3,536	2,019	2,051	3,024	3,615
b Business construction	2,762	3,129	2,186	2,783	3,300	3,513	4,062	4,366	4,477	4,385	4,581	3,800	2,232	1,097	936	1,180	1,461
c Net change in business inventories	+4,132	+7,375	+54	+534	+3,016	−917	+1,788	+1,586	+464	−321	+2,414	−1,128	−1,375	−2,461	−1,129	−1,524	+19
3 Public agencies	1,166	1,672	2,453	2,378	2,272	2,528	2,444	2,568	2,676	2,696	3,073	3,334	2,483	2,008	1,720	3,791	4,858
a Public construction	1,422	1,714	1,678	2,076	1,921	2,264	2,546	2,470	2,786	2,932	2,928	3,023	2,615	1,955	1,902	2,726	2,684
b Change in stocks of gold and silver	−256	−42	+775	+302	+351	+264	−102	+98	−110	−236	+145	+311	−132	+53	−182	+1,065	+2,174
4 Unallocable − net change in claims against foreign countries	+3,315	+2,254	+628	+215	−78	+446	+428	+44	+606	+957	+312	+371	+326	+40	+298	−868	−1,868
5 Gross capital formation, Variant I	19,341	22,100	11,488	13,282	18,199	15,245	19,211	19,037	18,208	17,824	20,298	13,662	8,464	3,147	4,268	6,061	9,008
6 Flow of consumers' durable commodities	5,987	6,921	5,570	6,181	7,943	7,900	9,056	9,445	8,890	9,174	9,913	7,550	5,748	3,806	3,882	4,686	5,918
7 Gross capital formation, Variant II	25,328	29,021	17,058	19,463	26,142	23,145	28,267	28,482	27,098	26,998	30,211	21,212	14,212	6,953	8,150	10,747	14,926
							1929 Prices										
Destined for use by																	
1 Consumers																	
a Residential construction	1,664	1,135	2,230	3,797	4,248	4,589	5,218	4,757	4,515	4,268	3,010	1,865	1,506	600	548	594	1,198
2 Business	10,332	11,534	5,382	7,079	11,037	7,276	11,038	11,306	10,902	10,118	13,886	8,544	5,052	668	2,155	2,977	5,988
a Flow of producers' durable commodities	4,633	4,735	3,303	3,858	5,058	4,838	5,368	5,761	5,993	6,083	6,891	5,791	4,012	2,601	2,779	3,714	4,312
b Business construction	2,776	2,476	2,221	2,973	3,186	3,408	4,026	4,325	4,467	4,391	4,581	3,884	2,481	1,332	1,166	1,403	1,741
c Net change in business inventories	+2,923	+4,323	−142	+248	+2,793	−970	+1,644	+1,220	+442	−356	+2,414	−1,131	−1,441	−3,265	−1,790	−2,140	−65
3 Public agencies	1,231	1,226	2,487	2,315	2,010	2,284	2,245	2,403	2,531	2,635	3,073	3,432	2,768	2,494	2,044	3,654	4,362
a Public construction	1,439	1,275	1,719	2,020	1,666	2,022	2,347	2,306	2,641	2,871	2,928	3,120	2,899	2,440	2,222	2,950	2,963
b Change in stocks of silver and gold	−208	−49	+768	+295	+344	+262	−102	+97	−110	−236	+145	+312	−121	+54	−178	+704	+1,399
4 Unallocable − net change in claims against foreign countries	+2,280	+1,391	+613	+212	−74	+433	+394	+42	+605	+943	+312	+409	+426	+59	+431	−1,104	−2,226
5 Gross capital formation, Variant I	15,507	15,286	10,712	13,403	17,221	14,582	18,895	18,508	18,553	17,964	20,281	14,250	9,752	3,821	5,178	6,121	9,322
6 Flow of consumers' durable commodities	5,821	5,707	4,580	5,819	7,522	7,873	8,817	9,752	9,364	9,555	9,894	7,875	6,577	4,704	4,645	5,259	6,756
7 Gross capital formation, Variant II	21,328	20,993	15,292	19,222	24,743	22,455	27,712	28,260	27,917	27,519	30,175	22,125	16,329	8,525	9,823	11,380	16,078

idential construction by David Wickens (see *Bulletin 65*, National Bureau of Economic Research, September 15, 1937), and upon a careful utilization of estimates and primary data available for nonresidential and farm construction. But the available data enable us to measure the value of construction at cost when completed by the construction firm, rather than at cost to the ultimate owner. The change in inventories is the least inclusive item, owing largely to scarcity of comprehensive data. The estimates cover most business inventories, include farm stocks for only three important crops and such livestock as is classified as non-durable, and stocks of gold and silver; but do not cover inventories of unincorporated business establishments in the service, finance and public utility groups, or inventories of non-business enterprises that are exempt from corporate income taxes. Moreover, this item is *not* the change from one year-end inventory to the next, obtained as the difference between successive year-end inventories in changing, current valuation. On the contrary, it is so measured as to reflect actual accretion of commodities to or actual drafts of commodities from the commodity stock comprising the inventories, this effect being attained by converting the inventories at each year-end to the commodity equivalent in constant prices, obtaining the change by subtraction, and then expressing the net change for each year in prices current during it. Finally, the balance of foreign trade in commodities is obtained directly from *The Balance of International Payments of the United States* published annually (recently semi-annually) by the Department of Commerce, and is as complete as foreign trade statistics allow.

The distribution by type of user also calls for some explanatory comment. First, our allocation of movable durable commodities between consumers' and producers' is crude, being based upon the characteristics of preponderant use, and the crudities influence both the volume of capital formation in Variant I and its distribution by type of user. No attempt has been made to segregate the shares of either consumers' or producers' durable commodities that flow to governmental agencies; or the shares of consumers' goods, such as passenger cars, that may be used by business enterprises; or the shares of producers' goods, such as typewriters or airplanes, that may be used by ultimate consumers. The result is to underestimate the volume of commodities destined for use by governmental agencies, to underestimate somewhat the volume of producers' goods, to underestimate gross capital formation in Variant I and to overestimate somewhat the volume of consumers' durable commodities. It is impossible to set even a tolerably approximate figure on these respective shortages and excesses. In percentages of such magnitudes as the total flow of consumers' or producers' durable commodities (including construction), they can hardly be significantly large. But this qualification is to be kept in mind in interpreting the totals in Table 10.

Second, changes in stocks of monetary metals (whether in bullion or the bullion contents of coinage) were classified under capital formation destined for use by public agencies. Since before 1933 part of that stock was held by private institutions, viz., banks, there may be some question concerning this classification. But it appeared to us that monetary metals, in bullion and coinage (excluding any business inventories held for industrial use) are largely in the nature of public capital, and should be classified in the same division as public roads, streets and governmental buildings.

The estimates in Table 10, as in all the subsequent basic tables, carry through 1935, but the measures for the last two years are built upon a relatively slender foundation and hence are much more tentative than those for the earlier years. Also, the estimates are in terms of both current market and 1929 prices. The adjustment for price changes was accomplished with the help of price indexes specially compiled for the purpose and based primarily upon the Bureau of Labor Statistics data on wholesale prices, scattered sample data on prices of machinery and on cost of construction, and information on the movement of transportation charges and distributive margins. The 1929 price level was chosen to express the volumes adjusted for price changes largely because, owing to the wealth of Census data for that year, the estimates for 1929 provided the best basing point for the entire inquiry. However, the general price level in 1929 was close to the average for the entire period and fairly close to the level of 1923–29.

2 VOLUME AND COMPOSITION OF GROSS CAPITAL FORMATION

Exclusive of consumers' durable commodities but inclusive of residential construction, gross capital formation in current prices averaged 14.0 billion; in 1929 prices 13.5 billion dollars (Table 11). The inclusion of movable consumers' commodities of durable character in Variant II of gross capital

Table 11
AVERAGE VOLUME AND DISTRIBUTION OF GROSS CAPITAL FORMATION

Destined for use by	AVERAGE VOLUME (millions of dollars) 1919–1935	PERCENTAGE DISTRIBUTION	PERCENTAGE DISTRIBUTION 1919–1927	1927–1935	1921–1927	1927–1933
			Current Prices			
1 Consumers						
a Residential construction	2,656	18.9	20.9	16.9	25.6	18.3
2 Business (total)	8,361	59.5	61.2	56.5	57.3	57.4
a Flow of producers' durable commodities	4,668	33.2	30.0	37.6	30.1	36.5
b Business construction	2,956	21.0	19.6	23.9	21.5	25.1
c Net change in business inventories	737	5.2	11.6	−5.0	5.7	−4.1
3 Public agencies (total)	2,595	18.5	12.9	26.4	15.1	21.0
a Public construction	2,332	16.6	12.1	23.3	13.7	21.1
b Change in stocks of silver and gold	263	1.9	0.8	3.1	1.4	−0.2
4 Unallocable—net change in claims against foreign countries	437	3.1	5.0	0.2	2.0	3.4
5 Gross capital formation, Variant I	14,050	100.0	100.0	100.0	100.0	100.0
6 Flow of consumers' durable commodities	6,975	49.6	43.5	59.0	48.0	57.0
7 Gross capital formation, Variant II	21,024	149.6	143.5	159.0	148.0	157.0
Destined for use by			1929 Prices			
1 Consumers						
a Residential construction	2,691	19.9	22.5	17.2	26.2	18.2
2 Business (total)	7,957	59.0	60.2	57.3	57.2	57.2
a Flow of producers' durable commodities	4,690	34.8	30.5	40.1	30.6	38.0
b Business construction	2,990	22.2	20.9	24.2	22.0	24.8
c Net change in business inventories	277	2.1	8.7	−7.0	4.7	−5.7
3 Public agencies (total)	2,541	18.8	13.1	25.6	14.5	21.1
a Public construction	2,343	17.4	12.2	23.8	13.2	21.3
b Change in stocks of silver and gold	198	1.5	0.9	1.9	1.4	−0.2
4 Unallocable—net change in claims against foreign countries	303	2.2	4.1	−0.1	2.0	3.5
5 Gross capital formation, Variant I	13,927	100.0	100.0	100.0	100.0	100.0
6 Flow of consumers' durable commodities	7,089	52.5	45.7	61.4	48.0	58.6
7 Gross capital formation, Variant II	20,581	152.5	145.7	161.4	148.0	158.6

formation raises the average volume to 21.0 billion dollars in current prices and to 20.6 billion in 1929 prices—an addition of some 50 per cent in current prices and 53 per cent in 1929 prices to the total of gross capital formation in Variant I. If the concept of capital formation is narrowed even further than in Variant I and confined to capital formation undertaken by business enterprises, the average volume in current prices shrinks to somewhat over 8 billion dollars, i.e., to 60 per cent of gross capital formation in Variant I and to 40 per cent of gross capital formation in Variant II.

In both variants presented in Tables 10 and 11 the volume of gross capital formation showed marked changes over the period. Variant I in current prices shows a decline from 22.1 billion in 1920 to 11.5 in 1921, over 48 per cent; a rise from 11.5 billion in 1921 to a peak of 20.3 billion in 1929, but interrupted by contractions in 1924 and 1927; and then a drastic contraction from 1929 to 1932, the volume in the latter year being only slightly over 15 per cent of the volume in the former year. The fluctuations in the volume measured in 1929 prices are somewhat less marked. But in both series for Variant I, the amplitude of expansions and contractions in gross capital formation was much greater than in the volume of gross national product of which this gross capital formation is a part.

In Variant I of gross capital formation, in which the only item destined for immediate use by consumers is residential construction, the most important division is that of volume destined for business use, whose average volume accounts for 60 per cent of the average of the total; and the most important single item is the flow of movable durable commodities to the producers who are their final users, which accounts for 33 per cent of the total. The relative share of public capital formation is on the average 18 per cent. While this is decidedly an underestimate, stemming from the crudity of our apportionment of durable commodity flow by type of user and from the failure to take account of inventories in the hands of public agencies, it may be doubted that more refined apportionment would result in a significant rise in the share in the total of public capital formation. The expansion of the concept of gross capital for-

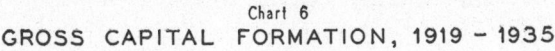

Chart 6
GROSS CAPITAL FORMATION, 1919 - 1935

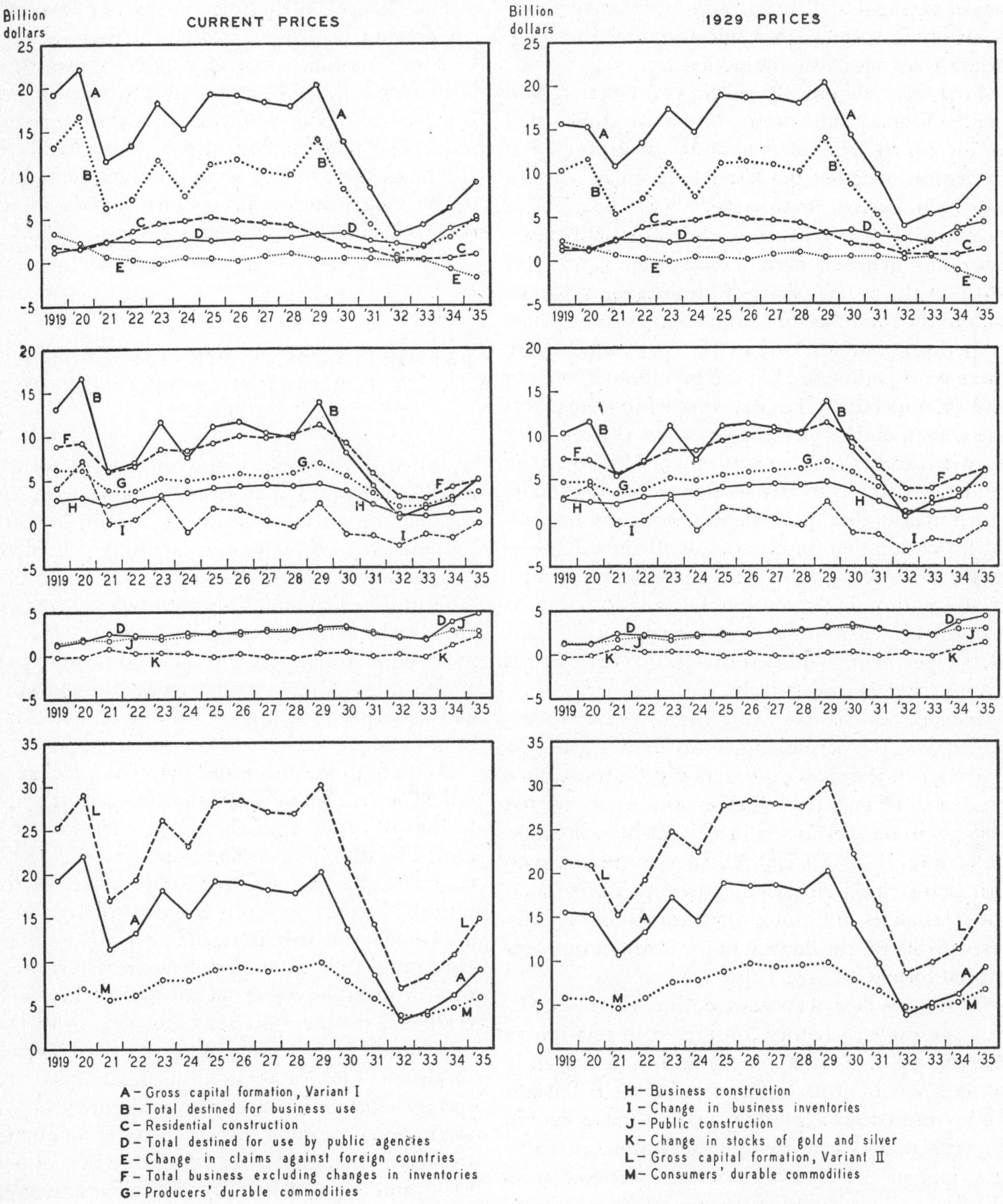
Billion dollars
CURRENT PRICES
1929 PRICES
A - Gross capital formation, Variant I
B - Total destined for business use
C - Residential construction
D - Total destined for use by public agencies
E - Change in claims against foreign countries
F - Total business excluding changes in inventories
G - Producers' durable commodities
H - Business construction
I - Change in business inventories
J - Public construction
K - Change in stocks of gold and silver
L - Gross capital formation, Variant II
M - Consumers' durable commodities

mation to include the flow of all consumers' durable finished commodities substantially changes the distribution by type of user. In the resulting Variant II of gross capital formation, it is the formation of capital destined for immediate use by consumers that accounts for the largest relative share (46 per cent); and the most important single item is the flow of consumers' durable movable commodities to their ultimate recipients, which accounts for 33 per cent of the new total.

The major items in gross capital formation show a marked long cyclical swing, both in absolute volume (see Chart 6) and in the percentage apportionment. Consequently only clearly marked and persistent changes in the relative distribution can be attributed significance as representing an approximation to long time movements. Such marked and persistent changes are few but the sets of averages in Table 11 and the relative movement of the volumes in Table 10 reveal three significant changes over the period.

First, the volume of public construction and hence of total public capital formation constituted an increasing share of the volume of gross capital formation, whether the latter is taken in Variant I or II, in current or 1929 prices (see the percentages in Table 11). Table 10 shows that this tendency was manifest even before 1929. Thus, for 1919–21 the average share of public capital formation in the total was for Variant I in current prices 10 per cent, and for Variant II, 7 per cent; for the three years ending in 1929 the percentages were 15 and 10, respectively. For the volumes in 1929 prices the corresponding percentages were 12 and 9 for 1919–21, and 15 and 10 for 1927–29. After 1929 the tendency became greatly accelerated owing to the much smaller decline in the volume of public capital formation than in the sum of all the other elements in the total. By 1935 the share of public capital formation in total capital formation was in Variant I in current prices 54 per cent; in Variant II, 33 per cent. The percentages in 1929 prices were 47 and 27, respectively.

Second, net changes in business inventories declined over the period and constituted a contracting share of the gross capital formation total. This tendency is manifest whether the ratio of net changes in business inventories is taken to the total in Variant I or Variant II, in current or 1929 prices. It is this decline in the percentage accounted for by changes in business inventories that is fully responsible for the decline in the share of business capital formation (see Table 11).

Third, the flow of consumers' durable commodities (movable), whether considered in relation to the volume of gross capital formation in Variant I or as a part of gross capital formation in Variant II, accounted for a growing share of gross capital formation. The movement of the volumes in Table 10 confirms the evidence of the percentages in Table 11. Thus, during 1919–21, in current prices, the flow of consumers' movable commodities averaged 35 per cent of total capital formation, Variant I; during 1927–29 the corresponding percentage was 50, and during 1933–35, 75. The movement of the percentages in terms of the larger volume of gross capital formation in Variant II likewise indicates the increasing relative importance of the flow of durable commodities to ultimate consumers, which again is confirmed by the volumes in 1929 prices.

The changes in the relative shares of the other components of gross capital formation appear to be either irregular, or if consistently shown by the percentages in Table 11, to lack significance. This is true even of the change in the relative share of residential construction which, though appreciable, appears to be due largely to the configuration of the long swing in its volume: a peak in 1925 rather than in 1929 appears to account for the changes in its percentage share indicated by Table 11.

3 APPORTIONMENT BETWEEN GROSS CAPITAL FORMATION AND CONSUMERS' OUTLAY

As stated in Section VI, the only variant of gross capital formation that is comparable with gross national product as measured in this report is Variant I, which includes residential construction but excludes the flow to ultimate consumers of all other finished commodities. But before gross capital formation so defined can be compared with gross national product and the volume of consumers' outlay obtained by subtraction, the conceptual and statistical comparability of the estimates must be assured.

By definition, the gross national product obtained as the sum of income payments and of gross savings of enterprises should equal the total obtained by adding consumers' outlay and gross capital formation. The only possible source of theoretical disparity between the two totals is in the adjustments made in the estimate of gross savings of enterprises in an attempt to have current consumption of durable capital goods and of inventories evaluated on their proper basis, i.e., on the basis of the market price current at the time income is produced. One of these adjustments may under certain conditions make for a disparity between the national product total that would be obtained by adding income payments and savings of enterprises and that obtained by adding consumers' outlay and capital formation.[21] But as the discussion in Appendix D shows, neither the adjustment for the difference between depreciation charges in cost and in current reproduction basis, nor that for gains and losses on inventory holding, as applied above, disturbs the theoretical comparability of the national product and the capital formation totals.

[21] For a more detailed discussion of this and other aspects of the comparison see Appendix D.

Table 12

APPORTIONMENT OF GROSS NATIONAL PRODUCT BETWEEN GROSS CAPITAL FORMATION AND CONSUMERS' OUTLAY, 1920-1934

(absolute figures in millions of dollars)

	1920	1921	1922	1923	1924	1925	1926	1927	1928	1929	1930	1931	1932	1933	1934
	\multicolumn{15}{c}{Current Prices}
1 Gross national product, 3-year moving average	72,578	72,057	70,516	74,730	80,139	83,661	86,324	88,537	90,157	88,805	80,371	64,892	52,830	49,835	54,515
2 Gross capital formation, Variant I, 3-year moving average	17,643	15,623	14,323	15,575	17,552	17,831	18,819	18,356	18,777	17,261	14,141	8,424	5,293	4,492	6,446
3 Consumers' outlay, line 1 - line 2	54,935	56,434	56,193	59,155	62,587	65,830	67,505	70,181	71,380	71,544	66,230	56,468	47,537	45,343	48,069
4 Gross capital formation as percentage of gross national product	24.3	21.7	20.3	20.8	21.9	21.3	21.8	20.7	20.8	19.4	17.6	13.0	10.0	9.0	11.8
5 Consumers' outlay as percentage of gross national product	75.7	78.3	79.7	79.2	78.1	78.7	78.2	79.3	79.2	80.6	82.4	87.0	90.0	91.0	88.2
	\multicolumn{15}{c}{1929 Prices}
1 Gross national product, 3-year moving average	64,471	65,973	69,481	74,722	79,170	82,154	84,660	87,440	89,860	89,554	83,715	71,926	63,796	62,554	67,562
2 Gross capital formation, Variant I, 3-year moving average	13,835	13,134	13,779	15,069	16,899	17,328	18,652	18,342	18,933	17,498	14,761	9,274	6,250	5,040	6,874
3 Consumers' outlay, line 1 - line 2	50,636	52,839	55,702	59,653	62,271	64,826	66,008	69,098	70,927	72,056	68,954	62,652	57,546	57,514	60,688
4 Gross capital formation as percentage of gross national product	21.5	19.9	19.8	20.2	21.3	21.1	22.0	21.0	21.1	19.5	17.6	12.9	9.8	8.1	10.2
5 Consumers' outlay as percentage of gross national product	78.5	80.1	80.2	79.8	78.7	78.9	78.0	79.0	78.9	80.5	82.4	87.1	90.2	91.9	89.8

But the conceptual similarity of the two measures does not change the effect on the comparison of the varying assumptions and approximations that had to be accepted in the statistical procedures leading to the final estimates. Since both sets of measures were built up part by part it would be an exceedingly laborious, and within the confines of this report an impossible, task to review the scaffolding by the help of which the final estimates have been erected. But even brief consideration of the most important deficiencies of the estimates in Appendix D suggests that the margins of error in the annual series in the comparison are such that it would be unwise to use the annual data and study the annual differences between the national product totals and those of capital formation. For this reason, three-year moving averages of these series have been calculated, centered on the middle year of the three; and the comparison is confined to the series thus smoothed.

We may now consider the apportionment of gross national product between gross capital formation and consumers' outlay (Table 12 and Chart 7). Table 12 reveals that during 1920–29 the volume of gross capital formation accounted for 20 per cent of the gross national product, the remaining four-fifths representing the outlay by consumers on commodities and services. In the contraction that followed 1929 the far more precipitous decline in the volume of capital formation than in consumers' outlay caused the ratio of the former to gross national product to fall to about half of what it had been before 1929. One aspect of the movement in this percentage apportionment deserves

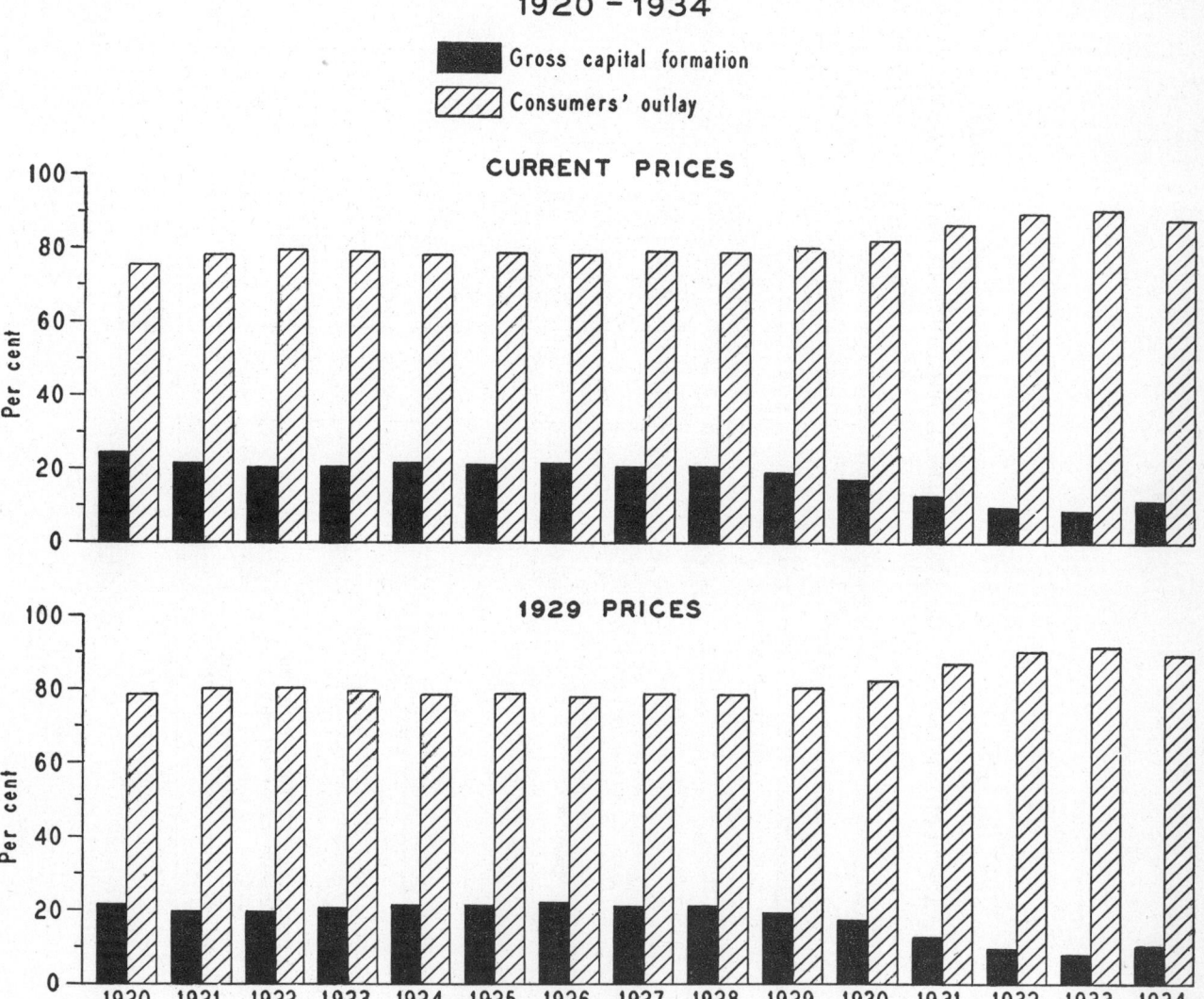

Chart 7
PERCENTAGE APPORTIONMENT OF GROSS NATIONAL PRODUCT BETWEEN GROSS CAPITAL FORMATION AND CONSUMERS' OUTLAY
1920 – 1934

comment, namely, its stability before 1929. On the assumption that since 1919 the volume of gross national product had described a complete long swing, one would expect to observe its reflection in an upward movement of the ratio of gross capital formation to gross national product from 1919, or the preceding or succeeding year, to 1929. For in all cyclical oscillations, especially of long duration, the volume of gross capital formation may be expected to rise more during the phase of expansion, just as it usually declines more during the phase of contraction. Gross capital formation did decline more after 1929, but its share did not increase before 1929. One reason for this stability may be looked for in the use of three-year moving averages; but their use has a relatively small effect since it did not conceal a marked decline after 1929. A more significant explanation may be that the long swing that culminated in 1929 may have begun before 1919. The War period, 1914–19, with its low volume of residential construction, was possibly characterized by a low ratio of gross capital formation to gross national product; although this surmise may be incorrect in view of the large net change in claims against foreign countries and extensive production of capital equipment during the War years. The trough of the ratio of gross capital formation to gross national product may have occurred before 1914, and the rise may have been from these low levels to the high plateau of 1920–29. Our study does not include years before 1919, but its results, shown in Table 12, suggest the importance of carrying the analysis back at least to the first decade of the twentieth century, if one is to understand clearly the developments since 1919.

VIII APPORTIONMENT OF NATIONAL INCOME BETWEEN NET CAPITAL FORMATION AND CONSUMERS' OUTLAY

1 NET CAPITAL FORMATION, VOLUME AND COMPOSITION

a *Characteristics of the Estimates*

THE volume of net capital formation is measured by subtracting from gross capital formation the estimated consumption of all durable capital goods utilized in the process of production. Such estimates have been prepared by Solomon Fabricant covering capital consumption: (a) that took place within the business enterprises of the nation (exclusive of that chargeable to residential buildings); (b) that was chargeable to the use of residential buildings; (c) that was chargeable to the use of durable goods by governmental agencies. Mr. Fabricant presented his preliminary results in *Bulletin 60*, Measures of Capital Consumption, 1919–1933, and we have taken advantage of the results of his subsequent work. Lack of data on the consumption of consumers' durable products except residential buildings and passenger cars made it impossible to measure net capital formation in Variant II; we had to confine our measures to net capital formation as described in Variant I (see Table 13 and Chart 8).

For the most important item in Table 13, capital formation destined for business use, there is some lack of correspondence between the gross capital formation totals (Table 10) and the totals of depreciation, depletion and fire loss deductions which are presented as measures of capital consumption. This lack of correspondence arises largely from two factors: (1) our distinction between producers' and consumers' goods is based on the preponderant use of the commodities, whereas the measures of depreciation, depletion, etc., charged by business enterprises are based on the actual segregation of capital goods used by business enterprises; (2) depreciation may be deducted for items not appearing in gross capital formation. Thus, the estimates in Table 13 of net capital formation destined for business use may be too large because: (a) gross capital formation totals include some durable goods that are destined for use either by ultimate consumers or by non-business agencies (e.g., government); (b) these totals include commodities (e.g., tools), which, their unit cost being small, may be treated by business enterprises on an inventory basis in 'deferred charges', rather than made subject to depreciation charges; (c) depreciation may be applied to capital values that have been reduced from their original cost. On the other hand, the net capital formation totals in Table 13 may be too small because: (a) the gross capital

Table 13

NET CAPITAL FORMATION, 1919-1935

(millions of dollars)

	1919	1920	1921	1922	1923	1924	1925	1926	1927	1928	1929	1930	1931	1932	1933	1934	1935
							Current Prices										
I Estimated consumption of capital goods in																	
1 Business use	6,297	7,263	5,492	5,293	6,002	5,857	5,954	6,550	6,457	6,674	7,134	6,746	6,058	5,200	4,921	5,395	5,669
2 Residential real estate	2,082	2,666	1,894	1,781	2,056	2,100	2,133	2,239	2,338	2,407	2,480	2,431	2,087	1,826	1,752	1,873	1,853
3 Government use¹	445	521	419	406	450	465	480	514	554	575	602	606	597	548	582	648	686
II Net capital formation																	
1 Business																	
a Incl. net change in business inventories	6,831	9,418	674	1,872	5,581	1,701	5,183	5,118	3,945	3,242	6,769	1,406	-1,665	-4,545	-3,063	-2,715	-574
b Excl. net change in business inventories	2,699	2,043	620	1,338	2,565	2,618	3,395	3,532	3,481	3,563	4,355	2,534	-290	-2,084	-1,934	-1,191	-593
2 Residential real estate	-350	-1,173	347	1,743	2,366	2,613	3,069	2,518	2,186	1,848	530	-626	-825	-1,382	-1,360	-1,415	-930
3 Public agencies	721	1,151	2,034	1,972	1,822	2,063	1,964	2,054	2,122	2,121	2,471	2,728	1,886	1,460	1,138	3,143	4,172
III Net change in claims against foreign countries	+3,315	+2,254	+628	+215	-78	+446	+428	+44	+606	+957	+312	+371	+326	+40	+298	-868	-1,868
IV Total net capital formation																	
1 Incl. net change in business inventories	10,517	11,650	3,683	5,802	9,691	6,823	10,644	9,734	8,859	8,168	10,082	3,879	-278	-4,427	-2,987	-1,855	800
2 Excl. net change in business inventories	6,385	4,275	3,629	5,268	6,675	7,740	8,856	8,148	8,395	8,489	7,668	5,007	1,097	-1,966	-1,858	-331	781
							1929 Prices										
I Estimated consumption of capital goods in																	
1 Business use	5,732	5,708	5,506	5,821	5,932	5,909	6,112	6,671	6,557	6,818	7,134	7,084	6,951	6,533	6,315	6,464	6,584
2 Residential real estate	2,000	2,027	1,885	1,919	1,975	2,045	2,139	2,239	2,333	2,414	2,480	2,511	2,490	2,467	2,447	2,433	2,407
3 Government use¹	397	394	405	432	424	450	479	514	551	584	602	631	669	696	722	755	788
II Net capital formation																	
1 Business																	
a Incl. net change in business inventories	4,600	5,826	-124	1,258	5,105	1,367	4,926	4,635	4,345	3,300	6,752	1,460	-1,899	-5,865	-4,160	-3,487	-596
b Excl. net change in business inventories	1,677	1,503	18	1,010	2,312	2,337	3,282	3,415	3,903	3,656	4,338	2,591	-458	-2,600	-2,370	-1,347	-531
2 Residential real estate	-336	-892	345	1,878	2,273	2,544	3,079	2,518	2,182	1,854	530	-646	-984	-1,867	-1,899	-1,839	-1,209
3 Public agencies	834	832	2,082	1,883	1,586	1,834	1,766	1,889	1,980	2,051	2,471	2,801	2,099	1,798	1,322	2,899	3,574
III Net change in claims against foreign countries	+2,280	+1,391	+613	+212	-74	+433	+394	+42	+605	+943	+312	+409	+426	+59	+431	-1,104	-2,226
IV Total net capital formation																	
1 Incl. net change in business inventories	7,378	7,157	2,916	5,231	8,890	6,178	10,165	9,084	9,112	8,148	10,065	4,024	-358	-5,875	-4,306	-3,531	-457
2 Excl. net change in business inventories	4,455	2,834	3,058	4,983	6,097	7,148	8,521	7,864	8,670	8,504	7,651	5,155	1,083	-2,610	-2,516	-1,391	-392

¹No depreciation charges were calculated on roads and sewers.

NET CAPITAL FORMATION

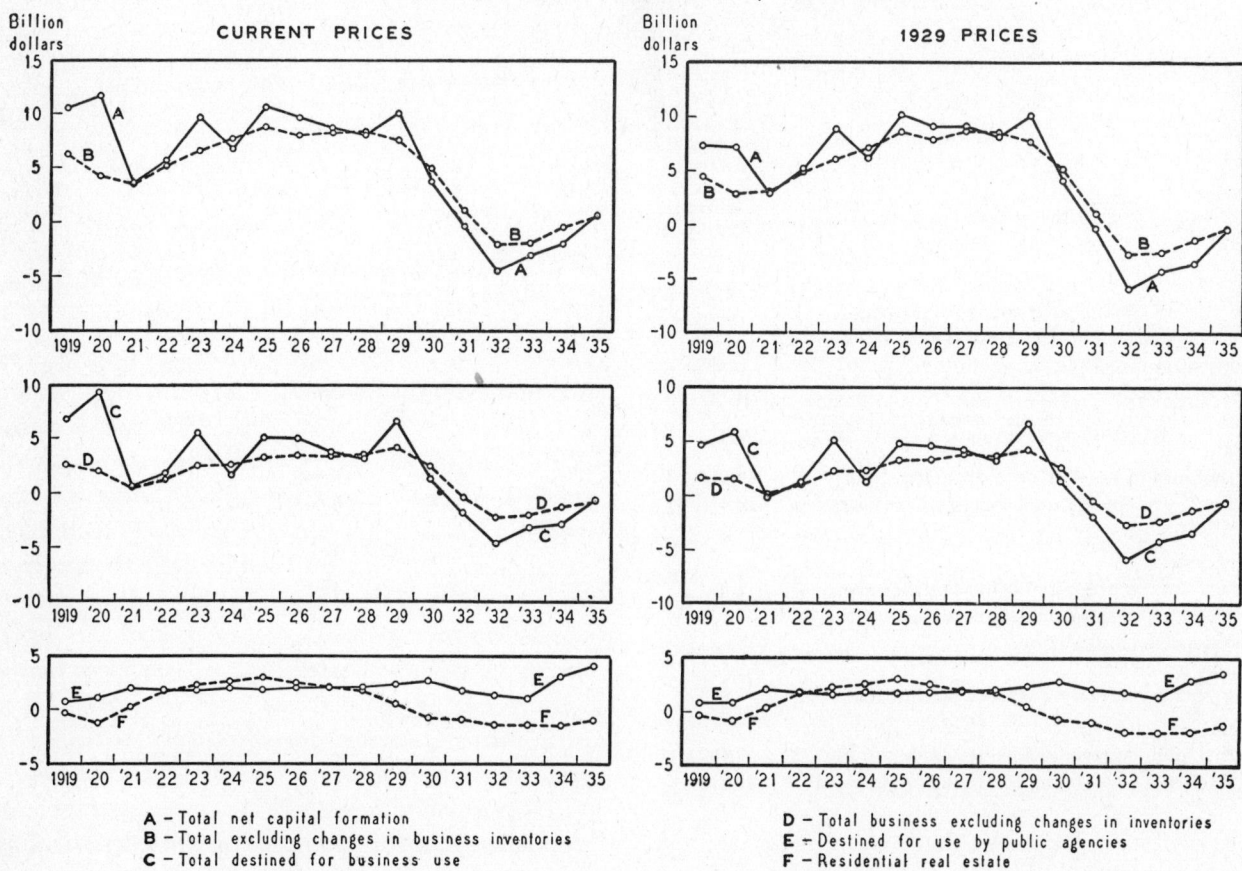

Chart 8
NET CAPITAL FORMATION, 1919-1935

A – Total net capital formation
B – Total excluding changes in business inventories
C – Total destined for business use
D – Total business excluding changes in inventories
E – Destined for use by public agencies
F – Residential real estate

formation total used fails to include some portion of commodities, classified by us as consumers' durable, that may be utilized by business enterprises (e.g., rugs); (b) depreciation charges may be applied to capital values that have become appreciated as compared with their original cost; or to intangibles; or to material repairs and alterations not included under gross capital formation in Table 13; or to commodities classified by us as consumers' durable but actually used by business enterprises. Similarly, the estimates of capital consumption for residential buildings, and especially for the durable commodities used by governmental agencies, are rough approximations. Obviously the measures in Table 13 are not of a high order of precision and should be used as approximations rather than as exact measures of net changes in the stock of capital goods held by the groups of users distinguished.

Nevertheless, the broad indications of the estimates are fairly trustworthy. We first compare them with gross capital formation and capital consumption, then discuss the absolute magnitude of net capital formation, and its distribution among the various components.

b *Comparison with Gross Capital Formation and Capital Consumption*

Comparison of gross capital formation, capital consumption and net capital formation shows what a large part of the total diverted into investment is offset by the current consumption of already existing durable commodities (Table 14). Of the average volume of gross capital formation for the entire period 62 per cent in current prices and 68 per cent in 1929 prices is accounted for by capital consumption, and only 38 and 32 per cent, respectively, can be considered as a net addition to the stock of capital goods.

This average percentage distribution of gross capital formation between presumptive replacement of capital consumed and net addition to stock varies little as between estimates in current and in 1929 prices; but it does vary significantly among the various groups distinguished in Tables 13 and 14. The relative share of capital consumption is greatest in residential construction, owing obviously to the existence of a large stock of residential buildings as compared with the moderate rate of gross additions to it during the period. In capital formation destined for business use the impor-

[49]

Table 14

AVERAGE VOLUME OF GROSS CAPITAL FORMATION, CAPITAL CONSUMPTION AND NET CAPITAL FORMATION

	GROSS CAPITAL FORMATION, AVERAGE VOLUME PER YEAR (millions of dollars)	CAPITAL CONSUMPTION		NET CAPITAL FORMATION		
		VOLUME PER YEAR (millions of dollars)	PER- CENTAGE OF GROSS CAPITAL FORMATION	VOLUME PER YEAR (millions of dollars)	PER- CENTAGE OF GROSS CAPITAL FORMATION	PERCENT- AGE OF NET CAPITAL FORMATION (INCLU- SIVE TOTAL, LINE 5a)
			Current Prices			
1 Business						
a Incl. net change in business inventories	8,361	6,057	72.4	2,305	27.6	43.2
b Excl. net change in business inventories	7,624	6,057	79.4	1,568	20.6	29.4
2 Residential construction	2,656	2,118	79.7	539	20.3	10.1
3 Public agencies	2,595	535	20.6	2,060	79.4	38.6
4 Net change in claims against foreign countries	437	0	0	437	100.0	8.2
5 Total						
a Incl. net change in business inventories	14,050	8,709	62.0	5,340	38.0	100.0
b Excl. net change in business inventories	13,313	8,709	65.4	4,603	34.6	86.2
			1929 Prices			
1 Business						
a Incl. net change in business inventories	7,957	6,343	79.7	1,614	20.3	37.2
b Excl. net change in business inventories	7,680	6,343	82.6	1,337	17.4	30.8
2 Residential construction	2,691	2,248	83.5	443	16.5	10.2
3 Public agencies	2,541	558	22.0	1,982	78.0	45.6
4 Net change in claims against foreign countries	303	0	0	303	100.0	7.0
5 Total						
a Incl. net change in business inventories	13,492	9,149	67.8	4,342	32.2	100.0
b Excl. net change in business inventories	13,215	9,149	69.2	4,066	30.8	93.6

tance of the replacement share is only slightly less, if we exclude changes in business inventories, a net item not subject to capital depreciation. Its inclusion serves to reduce the relative share of capital consumption in the total of gross capital formation destined for business use, and makes this share significantly lower than in residential construction. The apportionment between replacement and net additions is strikingly different, however, in the volume destined for use by governmental agencies. In the latter, capital consumption accounts for only one-fifth of the average volume of gross capital formation. The measure of capital consumption by governmental agencies is admittedly crude and incomplete, in that depreciation on roads and sewers is not allowed for. But this omission is perhaps justified on the ground that in these public properties little capital depreciation really occurs, capital consumed being replaced through repairs and maintenance. And whatever may be said of the possible underestimate of public capital consumption in the estimate presented, its share in the gross capital formation destined for public use may reasonably be expected to be very much lower than in residential construction or business capital, because the stock of public capital must have been small compared with the substantial gross additions since 1919 and because the rate at which the existing durable commodities in the hands of the government would depreciate would be extremely low.[22]

c Absolute Volume Compared with Wealth

The absolute volume of as inclusive a total of net capital formation as can be obtained with the available data is on the average 5.3 billion dollars per year in current prices, and 4.3 billion in 1929 prices. The significance of these figures is, perhaps, better comprehended when they are expressed in cumulative totals. If we add the net additions to the stock of capital goods that resulted during 1919–35 from the flow of producers' durable commodities to their ultimate domestic recipients, the volume of all new construction, net changes in business inventories and stocks of gold and silver, and net changes in claims against foreign countries, then, with each annual addition in current prices, the total amounts to 90.8 billion dollars; with these additions in 1929 prices, the total amounts to 73.8 billion. All this capital accumulation took place before the recent depression. The corresponding totals for the eleven years 1919–29 are, in current prices, 95.7 billion, in 1929 prices, 84.3 billion; and from 1930 through 1935 the net total added to the stock of capital goods was re-

[22] See also Section VIII, 1, d.

duced 4.9 billion dollars in current prices, 10.5 billion in 1929 prices.

It is of interest to compare this total of capital formation with the stock of wealth, to which it was an addition. The latest acceptable estimate of national wealth for this country is that made as of December 31, 1922 by the Federal Trade Commission (see *National Wealth and Income,* Washington, 1926). According to this report, total wealth at the current valuation amounted at the end of 1922 to 353 billion dollars. This total includes, however, 39.8 billion of furniture and personal effects, 4.6 billion of motor vehicles (Table 1, p. 28), and 122.2 billion of land exclusive of improvements (Table 3, p. 34). The first and last items should be completely omitted from the stock of wealth to which net capital formation, as measured in Table 13, could contribute; and the same is true of the preponderant portion of the value of motor vehicles. If, accordingly, we subtract 165 billion dollars from the total, the value of man-made wealth (excluding consumers' goods but including residential buildings) is some 188 billion dollars, at the end of and at valuation of 1922. If we assume that the same wealth would not be greatly different from the amount indicated if revalued at 1929 prices, an assumption whose arbitrariness is perhaps reduced by the fact that the general commodity price level in the two years is approximately the same, we can make the desired comparisons. The wealth at the beginning of 1919 must have amounted, in 1929 prices, to 188 billion minus the sum of net capital formation for 1919-22, i.e., minus 22.7 billion, or 165.3 billion. Hence the cumulative addition to this stock of capital goods during the seventeen years, 1919-35, amounted to about 44 per cent; and at this arithmetic rate of increase, the stock would have been doubled in about forty years. The total increase before 1930, i.e., before the depression, amounted, however, to 51 per cent of the stock at the beginning of 1919; and at the pre-depression rate, the stock would have doubled in twenty-two years. Whether either of the rates of capital accumulation thus shown appears high, average or low is hard to say, because our knowledge of capital growth in the past, the only basis for judgment, is so scanty.

d *Distribution Among Component Elements*

In considering the apportionment of total net capital formation among the distinguishable categories by type of user, the most striking feature is the relatively large amount destined for use by public agencies. Even if we disregard the rise in this item in 1934 and 1935, due largely to the influx of gold, the average volume for the period is not much below that of business net capital formation, and accounts for 30 per cent of the total. As will be seen from Table 14, this distribution of *net* capital formation makes a striking contrast to that of gross; when gross volumes are considered, capital formation destined for use by public agencies is less than one-third as large as that destined for use by business, and accounts for only 18 per cent of the total. The explanation lies in the materially smaller allowance for consumption of durable capital goods used by public agencies than of goods used by business firms or embodied in residential real estate.

The reasonableness of this difference among the various categories of capital goods with respect to the magnitude of the allowance for consumption and the resulting shift in the distribution from gross to net volumes has already been commented on and may be supported further by a brief inspection of the 1922 estimate of national wealth already referred to. According to this estimate, the value of improvements embodied in streets, roads and other highway structures not covered under exempt real estate, and in exempt real estate amounted to 20.8 billion dollars (*National Wealth and Income,* Table 3). This left some 167 billion as the value of man-made wealth (i.e., excluding land) in use by business agencies or embodied in residential structures. Table 14 shows that the average volume of gross capital formation for use by public agencies was about 2.6 billion dollars, whereas that for use by business or resulting from residential construction amounted to about 11.0 billion. Thus even the gross additions were at a higher relative rate for capital destined for use by public agencies than for business or residential construction. If, furthermore, we take into consideration the naturally much lower rate of capital consumption of goods in public use, the results in Table 14 are easily comprehended.

However, gross rather than net capital formation provides the proper guide to the relative importance of the various categories of capital goods in the economic life and industrial structure of the nation. The line between replacement demand and expansion demand for capital goods is thin and tenuous; and it is the total volume that controls the relative importance of a given category of capital goods and of the changes in their flow. True, the relatively large share of public agencies in net capital formation, if continued, will eventually modify greatly the structure of national wealth, and perhaps also the structure of the current production of capital goods. But for the present, it is gross capital formation, with its materially different distribution among business, ultimate consumers and public agencies that provides the more valid notion of the relative importance of various capital goods categories in the functioning of the economic system.

Finally, it should be noted that the movements in the percentage distribution of net capital formation among its component elements parallels those observed for gross capital formation, Variant I. The volume of public construction and hence of public capital formation accounts for an increasing percentage of total net capital formation; whereas the net changes in business inventories account for an algebraically diminishing proportion. These shifts in favor of public capital investment, and within business capital formation from investment in inventories to investment in capital equipment, are characteristic of the period since 1919, and even more marked for net capital formation than for gross.

2 APPORTIONMENT OF NATIONAL INCOME BETWEEN NET CAPITAL FORMATION AND CONSUMERS' OUTLAY

National income, as defined and measured above, is directly comparable with net capital formation; and the volume of consumers' outlay can again be obtained by subtracting net capital formation from national income. But differences in the assumptions underlying the estimates of the two totals necessitate the use of three-year moving averages rather than values for single years (Table 15 and Chart 9).

The first interesting conclusion suggested by Table 15 concerns the relatively small proportion that net additions to the stock of capital goods, as measured by us, constitute of total national income. The average share over the period is about 8 per cent, in contrast to the share of the comparable measure of gross capital formation in gross national product of about 19 per cent. This difference in the percentage distribution is obviously due to the subtraction, in arriving at net capital formation, of all consumption of the stock of capital goods from the gross volume. It is thus seen that of the total *net* output of commodities and services only a relatively small fraction, even in the most prosperous years, can be characterized as net addition to the stock of capital goods. Even during prosperous years over 87 per cent of the current output is in the group of immediately consumed commodities and services.

This relatively small share of the net product that constitutes a net addition to the nation's stock of capital goods fluctuates violently over the period. Even with the short term fluctuations smoothed out by the application of a three-year moving average, it almost doubles from 1921 to 1926, when computed for volumes in current prices; and increases almost a half from 1921 to 1926, when computed for volumes in 1929 prices. Its decline after 1928–29 is, of course, still more marked. And instead of the stability during 1920–29 in the ratios of capital formation to national product, observed in the comparison of gross volumes, there is a definite upward movement to 1924 or 1926 in the ratios in Table 15.

The volume of consumers' outlay, in contrast to the volume and share of net capital formation, shows no marked fluctuations, especially when the effect of changing price levels is removed. When measured in constant prices and in terms of a three-year moving average, it does not decline until 1930; and while the subsequent contraction to 1932 is fairly substantial, its movement over the period as a whole is distinctly upward. This contrast in movement and variability between consumers' outlay and capital formation clearly justifies the emphasis that economic science places upon the distinction between capital goods and consumable goods; and renders it important to provide separate and comprehensive measures of the volume of consumers' outlay and of capital formation as the basis for a further study of the various economic forces that operate to produce divergent movements in these two, essentially related, segments of the national product.

Table 15

APPORTIONMENT OF NATIONAL INCOME BETWEEN NET CAPITAL FORMATION AND CONSUMERS' OUTLAY, 1920-1934

(absolute figures in millions of dollars)

	1920	1921	1922	1923	1924	1925	1926	1927	1928	1929	1930	1931	1932	1933	1934
						Current Prices									
1 National income, 3-year moving average	63,552	63,479	62,585	66,594	71,640	74,897	77,251	79,101	80,417	78,920	70,791	56,193	44,974	42,253	46,722
2 Net capital formation, inclusive total, 3-year moving average	8,617	7,045	6,392	7,439	9,053	9,067	9,746	8,920	9,036	7,376	4,561	-275	-2,564	-3,090	-1,347
3 Consumers' outlay, line 1 - line 2	54,935	56,434	56,193	59,155	62,587	65,830	67,505	70,181	71,381	71,544	66,230	56,468	47,538	45,343	48,069
4 Net capital formation as percentage of national income	13.6	11.1	10.2	11.2	12.6	12.1	12.6	11.3	11.2	9.3	6.4	-0.5	-5.7	-7.3	-2.9
5 Consumers' outlay as percentage of national income	86.4	88.9	89.8	88.8	87.4	87.9	87.4	88.7	88.8	90.7	93.6	100.5	105.7	107.3	102.9
						1929 Prices									
1 National income, 3-year moving average	56,453	57,941	61,381	66,419	70,682	73,301	75,462	77,880	80,036	79,468	73,531	61,915	54,032	52,943	57,924
2 Net capital formation, inclusive total, 3-year moving average	5,817	5,101	5,679	6,766	8,411	8,476	9,454	8,781	9,108	7,412	4,577	-736	-3,513	-4,571	-2,765
3 Consumers' outlay, line 1 - line 2	50,636	52,840	55,702	59,653	62,271	64,825	66,008	69,099	70,928	72,056	68,954	62,651	57,545	57,514	60,689
4 Net capital formation as percentage of national income	10.3	8.8	9.3	10.2	11.9	11.6	12.5	11.3	11.4	9.3	6.2	-1.2	-6.5	-8.6	-4.8
5 Consumers' outlay as percentage of national income	89.7	91.2	90.7	89.8	88.1	88.4	87.5	88.7	88.6	90.7	93.8	101.2	106.5	108.6	104.8

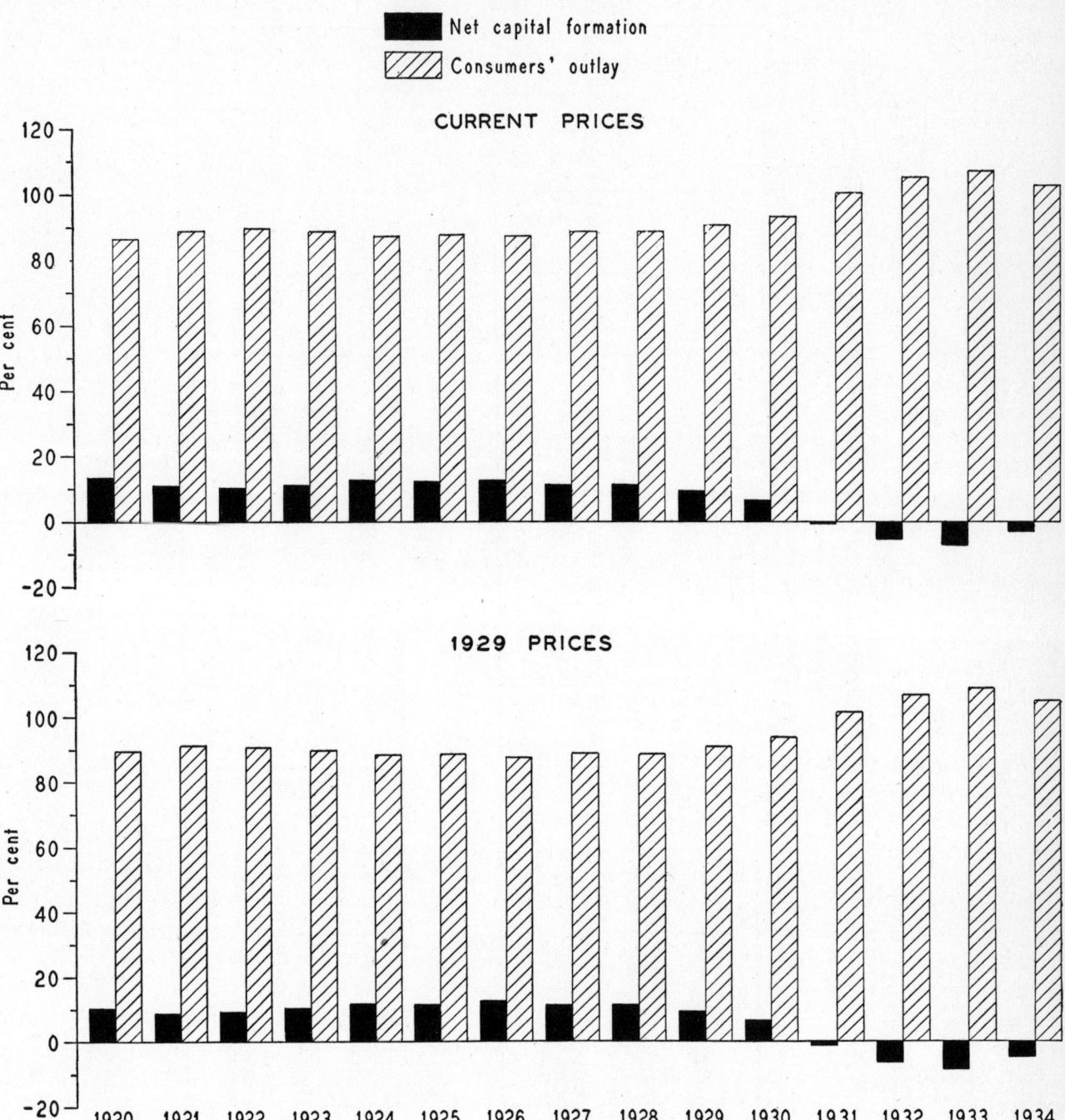

Chart 9
PERCENTAGE APPORTIONMENT OF NATIONAL INCOME BETWEEN NET CAPITAL FORMATION AND CONSUMERS' OUTLAY 1920-1934

IX THE COMPOSITION OF CONSUMERS' OUTLAY

THE composition of capital formation was discussed in Sections VII and VIII in connection with the apportionment of the national product between capital formation and consumers' outlay: the limitations of the statistical measures provided and the exact meaning of the apportionment could be grasped only if the characteristics of the estimates of capital formation were discussed and the contents of those totals analyzed. The measure of consumers' outlay was obtained not by direct estimate but by subtracting gross and net capital formation from gross and net national product respectively. The resulting volume of consumers' outlay is, in accordance with the definition, the same whether obtained as part of gross or of net national product.

The study of capital formation yielded measures of the flow not only of finished durable but also of finished perishable and semidurable commodities.[23] The totals for the latter commodity groups were gauged at the same stage of economic circulation and with the same attempt at complete coverage as the estimates of the flow of durable commodities. Hence the resulting totals represent the movement of all finished perishable and semidurable commodities to ultimate domestic consumers at cost to them; they omit only such minor commodity groups as flowers and the small volume of commodities produced in the service industries and thus not included under farming, mining or manufacturing.

In Variant I, consumers' outlay includes consumers' durable commodities, exclusive of residential construction. The three groups—perishable, semidurable and durable—together account for all the commodities whose cost is a part of consumers' outlay. The residual portion of consumers' outlay represents largely the cost to consumers of services not embodied in new commodities: those rendered in the form of repairs of consumers' durable commodities, including such repairs of residential buildings as do not call for building permits; those rendered largely by durable commodities to consumers (for example, by residential buildings to their tenants, transportation services to ultimate consumers by railroads); those rendered directly by individuals to individuals (for example, services of physicians and teachers). The difference between consumers' outlay and the flow of all finished commodities, being a residual item, reflects also whatever deficiencies are present in our measures of capital formation (Sec. VII, 1). But it is doubtful that these elements, which do not represent the cost to consumers of services not embodied in new commodities, affect substantially the magnitude of the difference between consumers' outlay and the flow of all finished commodities to their ultimate users.

The distribution of consumers' outlay among various commodity groups according to their durability, and among all new commodities and services not embodied in new commodities, is shown in Table 16 and Chart 10. Of the total outlay by consumers, the cost of perishable commodities accounts, on the average, for about 40 per cent; of semidurable commodities, for about 17 per cent; of durable commodities, exclusive of residential construction for 12 per cent; and of services not embodied in new commodities, for about 31 per cent.

In the distribution of consumers' outlay in current prices two significant movements appear. First, the relative share of semidurable commodities declines throughout the period: instead of accounting for slightly less than one-fifth of the total, as they do in 1920, they constitute only slightly more than one-seventh by the end of the period. Second, the relative share of consumers' durable commodities describes a substantial long swing, rising from 1921 to 1926 and declining from 1926 to 1933. The shares of the other two categories in the distribution appear relatively stable for the major part of the period covered. Thus the percentage accounted for by perishable commodities ranges from 38.3 to 39.4 for 1922 through 1931, the variations occurring within this limited range showing no consistency. Similarly, the share of services not embodied in new commodities ranges from 29.9 to 32.8 per cent for 1922 through 1929. Only in the disturbed years following the end of the War (1920–21) and during the recent depression did the share of these two categories change significantly.

The adjustment of the volume of consumers' outlay and of its component elements for price changes is subject to serious qualifications. While

[23] For a definition of these groups, see Section VI, 2.

Table 16

COMPOSITION OF CONSUMERS' OUTLAY, 1920-1934

Part A. Absolute Figures, Three-Year Moving Averages

(millions of dollars)

	1920	1921	1922	1923	1924	1925	1926	1927	1928	1929	1930	1931	1932	1933	1934
Current Prices															
1 Perishable commodities	24,657	23,578	22,141	22,709	24,040	25,420	26,394	27,042	27,523	27,431	25,475	22,008	19,254	19,012	20,661
2 Semidurable commodities	10,781	10,638	10,361	10,694	11,140	11,338	11,770	12,047	12,202	11,769	10,712	8,826	7,420	6,916	7,392
3 Consumers' durable commodities	6,159	6,224	6,565	7,341	8,300	8,800	9,130	9,170	9,326	8,879	7,737	5,701	4,479	4,125	4,829
4 Services not embodied in new commodities	13,338	15,994	17,126	18,411	19,107	20,272	20,211	21,922	22,330	23,465	22,306	19,933	16,385	15,290	15,187
5 Total consumers' outlay	54,935	56,434	56,193	59,155	62,587	65,830	67,505	70,181	71,381	71,544	66,230	56,468	47,538	45,343	48,069
1929 Prices															
1 Perishable commodities	21,075	22,007	22,870	24,045	24,915	25,883	26,412	26,985	27,603	27,871	27,716	27,001	26,786	26,862	26,820
2 Semidurable commodities	7,374	7,826	8,907	9,329	9,726	9,843	10,620	11,076	11,717	11,513	11,248	10,450	9,777	9,273	9,418
3 Consumers' durable commodities	5,369	5,369	5,974	7,071	8,071	8,814	9,311	9,557	9,604	9,108	8,115	6,385	5,309	4,869	5,553
4 Services not embodied in new commodities	16,818	17,638	17,951	19,208	19,559	20,285	19,665	21,481	22,004	23,564	21,875	18,815	15,673	16,510	18,898
5 Total consumers' outlay	50,636	52,840	55,702	59,653	62,271	64,825	66,008	69,099	70,928	72,056	68,954	62,651	57,545	57,514	60,689

Part B. Percentage Distribution

	1920	1921	1922	1923	1924	1925	1926	1927	1928	1929	1930	1931	1932	1933	1934
Current Prices															
1 Perishable commodities	44.9	41.8	39.4	38.4	38.4	38.6	39.1	38.5	38.6	38.3	38.5	39.0	40.5	41.9	43.0
2 Semidurable commodities	19.6	18.9	18.4	18.1	17.8	17.2	17.4	17.2	17.1	16.5	16.2	15.6	15.6	15.3	15.4
3 Consumers' durable commodities	11.2	11.0	11.7	12.4	13.3	13.4	13.5	13.1	13.1	12.4	11.7	10.1	9.4	9.1	10.0
4 Services not embodied in new commodities	24.3	28.3	30.5	31.1	30.5	30.8	29.9	31.2	31.3	32.8	33.7	35.3	34.5	33.7	31.6
5 Total consumers' outlay	100.0	100.0	100.0	100.0	100.0	100.0	100.0	100.0	100.0	100.0	100.0	100.0	100.0	100.0	100.0
1929 Prices															
1 Perishable commodities	41.6	41.6	41.1	40.3	40.0	39.9	40.0	39.1	38.9	38.7	40.2	43.1	46.5	46.7	44.2
2 Semidurable commodities	14.6	14.8	16.0	15.6	15.6	15.2	16.1	16.0	16.5	16.0	16.3	16.7	17.0	16.1	15.5
3 Consumers' durable commodities	10.6	10.2	10.7	11.9	13.0	13.6	14.1	13.8	13.5	12.6	11.8	10.2	9.2	8.5	9.1
4 Services not embodied in new commodities	33.2	33.4	32.2	32.2	31.4	31.3	29.8	31.1	31.0	32.7	31.7	30.0	27.2	28.7	31.1
5 Total consumers' outlay	100.0	100.0	100.0	100.0	100.0	100.0	100.0	100.0	100.0	100.0	100.0	100.0	100.0	100.0	100.0

the price data available for the perishable and semidurable commodities make possible a tolerably reliable adjustment for price changes in these two groups, the information for consumers' durable commodities is scanty and the adjustment for

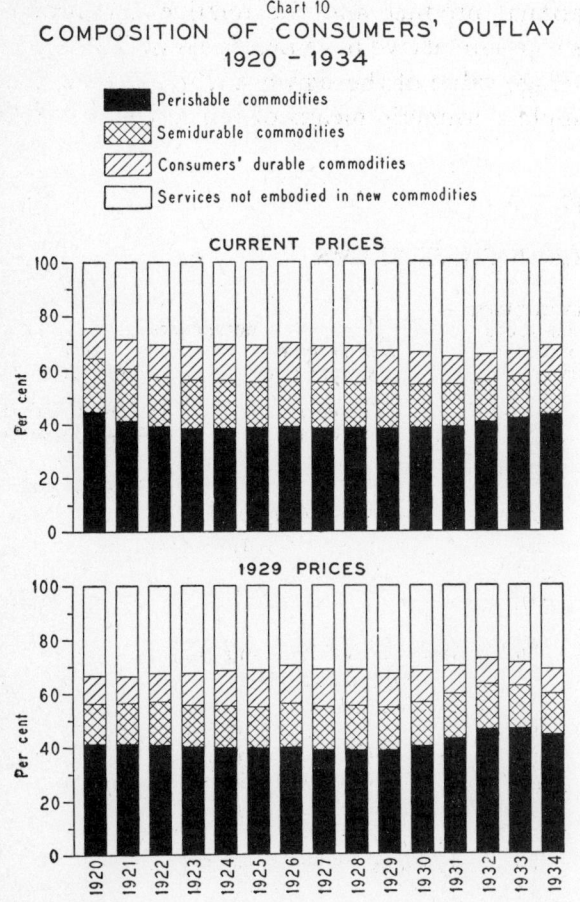

Chart 10
COMPOSITION OF CONSUMERS' OUTLAY
1920 – 1934

price changes correspondingly poor. For the group of services not embodied in new commodities, whose volume in constant prices is obtained by subtraction, the price adjustment obviously reflects the crudities and possible errors in the adjustment for price changes of national income, of net capital formation and of the three commodity groups. Any conclusions that may be derived from the apportionment of consumers' outlay in constant prices, are, to an even greater extent than those established for the distribution in current prices, tentative suggestions subject to further check.

In the distribution of consumers' outlay in 1929 prices, one significant movement stands out, namely, the substantial rise in the relative share of consumers' durable commodities from 1921 through 1926 and the marked decline from 1926 through 1933. While in its timing this long swing parallels the one observed for distribution in current prices, its amplitude is greater, owing primarily to the difference in the movement of prices as between consumers' durable commodities and other commodities and services, largely perishable commodities. From 1921 to 1926 the prices of durable commodities rose; the prices of perishable commodities declined. After 1926, especially after 1929, perishable commodities dropped much more precipitously in price than durable commodities. Thus the differences in price movement for the long swing ran counter to the movement in the relative proportion of quantities, and tended to damp the amplitude of the swing when changing quantities and prices were combined in the volumes at current prices.

For the relative shares of the other commodities, the distribution in 1929 prices shows a high degree of stability. Thus the percentage accounted for by perishable commodities ranges from 38.7 to 41.6 for 1920 through 1930; and that for services not embodied in new commodities from 29.8 to 32.7 for 1922 through 1931. For semidurable commodities the percentage share fluctuated between 15.2 and 16.5 for 1922 through 1930; and the disappearance of the downward movement in the relative share of this group in the distribution in current prices is obviously due to the smaller decline or greater rise in the prices of semidurable goods, primarily clothing, than in the prices of the other components in consumers' outlay.[24] Only during the disturbed years immediately after the War and during the recent depression have any marked changes occurred in the distribution of consumers' outlay measured in 1929 prices.

[24] This is confirmed by the fact that the price index for the clothing item in the Bureau of Labor Statistics cost of living index declined from 290.6 in 1920 (1913=100) to 132.0 in 1932; the total cost of living index, from 202.4 to 139.9.

X SUMMARY

This report is devoted primarily to measuring the volume of the national product, establishing the relative magnitudes of its significant parts, and describing the broad changes in both the total and its components.

To give the reader at a glance the volume of the national product and the relative magnitude of its components we have presented in Table 17 the average value of the significant items, obtained as simple arithmetic means of the respective annual

Table 17

AVERAGE VALUE OF SIGNIFICANT ITEMS

	AVERAGE VALUE (*millions of dollars*)	PERCENTAGES, SUBGROUPS OF RESPECTIVE TOTALS	REFERENCES TABLE PROVIDING ANNUAL DATA	TEXT DISCUSSION
1 Gross national product	73,107		Table 1	Sections I and II
2 National income	64,397		Table 1	Sections I and II
National income according to industrial origin [1]:				Section III
3 Agriculture	6,253	9.8	Appendix Table I	
4 Mining	1,562	2.4	" " II	
5 Manufacturing	14,008	21.9	" " "	
6 Construction	2,078	3.3	" " "	
7 Transportation and other public utilities				
a Electric light and power and manufactured gas	1,051	1.6	Appendix Tables I and III	
b Steam railroads, Pullman and express	3,737	5.8	Appendix Table I	
c Pipe lines, street railways, and water transportation	1,131	1.8	" " II	
d Communication	719	1.1	" " I	
e Total	6,638	10.4	" " II	
8 Trade	8,825	13.8	" " II	
9 Finance				
a Banking	884	1.4	Appendix Tables I and III	
b Insurance	861	1.3	" " " " "	
c Real estate	5,086	8.0	" " " " "	
d Total	6,831	10.7	Appendix Table II	
10 Government	7,036	11.0	" " I	
11 Service	8,404	13.1	" " II	
12 Miscellaneous	2,275	3.6	" " II	
13 Total national income	63,910	100.0	" " II	
National income according to type of income [2]:				Sections IV and V
14 Employees' compensation	44,666	70.0	Table 6	
15 Withdrawals by entrepreneurs	7,789	12.2	"	
16 Rents	2,964	4.6	"	
17 Entrepreneurial income payments	10,753	16.9	"	
18 Dividends	3,763	5.9	"	
19 Interest	4,494	7.0	"	
20 Property income payments	8,385	13.1	"	
21 Aggregate income payments to individuals	63,804	100.0	"	
22 Net savings of business enterprises	250	0.4	"	
23 Net savings of government	344	0.5	"	
24 Total net savings of enterprises	593	0.9	"	
25 National income	64,397	100.9	"	
26 Gross savings of enterprises	9,303	14.6	"	
27 Gross national product	73,107	114.6	"	

[1] Before adjustments for the effects of inventory revaluations, differences between depreciation and depletion at cost prices and current reproduction prices, and gains and losses on the sale of capital assets.
[2] After the adjustments mentioned in footnote 1.

SUMMARY

Table 17 (*continued*)

	AVERAGE VALUE (*millions of dollars*)	PERCENTAGES, SUBGROUPS OF RESPECTIVE TOTALS	REFERENCES TABLE PROVIDING ANNUAL DATA	TEXT DISCUSSION
Apportionment of gross national product between gross capital formation and consumers' outlay:				Sections VI and VII
28 Gross capital formation	14,050	19.2	Appendix Table VIII	
29 Consumers' outlay	59,057	80.8	" " "	
30 Gross national product	73,107	100.0	" " "	
Apportionment of national income between net capital formation and consumers' outlay:				Sections VI and VIII
31 Net capital formation	5,340	8.3	Appendix Table VIII	
32 Consumers' outlay	59,057	91.7	" " "	
33 National income	64,397	100.0	" " "	
Composition of gross capital formation:				Section VII
34 Consumers				
a Residential construction	2,656	18.9	Table 10	
35 Business				
a Flow of producers' durable commodities	4,668	33.2	"	
b Business construction	2,956	21.0	"	
c Net change in business inventories	737	5.2	"	
d Total	8,361	59.5	"	
36 Public agencies				
a Public construction	2,332	16.6	"	
b Change in stocks of silver and gold	263	1.9	"	
c Total	2,595	18.5	"	
37 Unallocable—net change in claims against foreign countries	437	3.1	"	
38 Gross capital formation, Variant I	14,050	100.0	"	
39 Flow of consumers' durable commodities	6,975	49.6	"	
Composition of net capital formation:				Section VIII
40 Business	2,305	43.2	Table 13	
41 Residential construction	539	10.1	"	
42 Public agencies	2,060	38.6	"	
43 Net change in claims against foreign countries	437	8.2	"	
44 Net capital formation	5,340	100.0	"	
Composition of consumers' outlay:				Section IX
45 Perishable commodities	23,834	40.4	Appendix Table VIII	
46 Semidurable commodities	10,174	17.2	" " "	
47 Consumers' durable commodities	6,975	11.8	" " "	
48 Services not embodied in new commodities	18,074	30.6	" " "	
49 Total consumers' outlay	59,057	100.0	" " "	

estimates, and the percentage distributions based on these average values. The tables in which the detailed annual estimates are given, and the section in the text where the scope of the measures, their limitations, and the changes they reveal are discussed are also indicated. Following the order suggested by Table 17 we now summarize the broad changes.

(1) The movements of the national product, which reflected fairly clearly the cyclical fluctuations in this country's economic activity, were accompanied by an increase in population and by a still greater increase in the number of gainfully occupied and of equivalent consuming units that this population represented. Hence, per capita income, and still more the income per gainfully occupied or per consuming unit, rose much less between 1921 and 1929, and declined more from 1929 to 1933, than did total national income.

(2) In the distribution in current prices the combined share of commodity producing industries (agriculture, mining, manufacturing, construc-

tion) in the national product declined over the period; that of commodity handling industries (transportation and other public utilities, trade) remained stable; that of service industries (finance, government, service, miscellaneous) rose. But in the distribution in 1929 prices the relative share of commodity producing industries and of the total group of 'other industries' remained stable. This suggests that the decline in the relative share of commodity producing industries, in the distribution in current prices, was due largely to a decline in commodity prices greater than in the prices of other consumers' goods; but this conclusion is highly tentative owing to the crudity of the adjustment for price changes.

(3) Gross, and especially net, savings of enterprises fluctuated much more violently than aggregate income payments. Hence, in spite of the small average share of net savings in national income, its fluctuations accounted for a major share in the variation in national income between good and bad years.

(4) The share of employees' compensation in aggregate income payments rose slightly over the period. But the rise was so small as to be insignificant, and study of the relative distribution, industry by industry, revealed that the proportion of employees' compensation declined in most industries. Thus the stability and slight rise in the percentage of aggregate income payments accounted for by employees' compensation was due exclusively to a shift of the industrial distribution in favor of those industrial branches that are marked by a high ratio of employees' compensation to total payments originating. The share of entrepreneurial income payments declined distinctly over the period, as a result of both the intra-industry decline in its relative share and the inter-industry shift in favor of industries with a low ratio of entrepreneurial income payments to total payments. The share of property income rose substantially over the period, owing largely to the increased proportion of property income payments within most of the industries.

(5) Consumers' outlay, taken as a part of either gross national product or of national income, was significantly more stable over time than the national product itself. The other share, gross or net capital formation, especially the latter, fluctuated conspicuously, becoming negative during some years of the recent depression.

(6) When the components of both gross and net capital formation were distinguished, it was seen that the relative share of public construction rose and that of net changes in business inventories declined over the period in both totals. When the flow of consumers' durable commodities, exclusive of residential construction, was considered, their importance, relative to the volume of gross capital formation in Variant I, appeared to have increased.

(7) In the distribution of consumers' outlay in current prices the share of semidurable commodities declined over the period, that of consumers' durable commodities described a long swing with a rise to 1926 and a decline to 1933, while the share of the other two categories was on the whole stable except immediately after the War, 1920–21, and during the recent depression, 1930–33. In the distribution in 1929 prices the only significant movement was a long swing in the share of consumers' durable commodities, a swing of even greater amplitude than in the distribution in current prices. The share of the other categories was fairly stable throughout the period except, again, for the disturbed years 1920–21 and 1930–33.

APPENDIX A

NATIONAL INCOME AND AGGREGATE INCOME PAYMENTS TO INDIVIDUALS BY TYPE OF INCOME AND INDUSTRIAL BRANCH, BASIC AND OTHER VARIANTS

APPENDIX A presents in detail the estimates upon which the summary tables and measures in the text are based, as well as other variants of these measures not considered basic and not utilized in the discussion.

The basic variant, so far as the available data admit of its measurement in strict conformity with the concepts established, is presented in Appendix Table I. In this variant, the aggregate of income payments to individuals is the sum of employees' compensation, entrepreneurial withdrawals, and individuals' receipts of net rents (paid and imputed), dividends and interest. Net business savings, as reported in the accounts of business concerns, are in this variant adjusted for: (a) gains and losses on the sale of capital assets; (b) that element of revaluation of inventories which is retained, under usual accounting procedures, in net profit or loss after payment of dividends; (c) the difference between depreciation and depletion at book value and at reproduction prices. However, gains and losses on the sale of capital assets could be excluded only since 1929; the adjustment of net savings for the effect of changing inventory valuation is not complete in any industrial branch, and could not be made even in its usual form in the subdivisions of the transportation and other public utilities and the finance groups; and the adjustment (c), an attempt to use a theoretically appropriate measure of capital consumption, could be made only for the economic system as a whole, not by industrial branches. For these reasons and in these respects, even the basic variant, presented in Appendix Table I, departs from the theoretical concept of national income.

Since opinions differ concerning the limits of such concepts as aggregate income payments to individuals and national income, and since the accounting measures of net business savings have an interest of their own, we present in Appendix Table II measures of other variants of total national income and of some of its components. These variants are of two distinct types. One is suggested by the general difficulty of distinguishing between entrepreneurial withdrawals and the net savings of entrepreneurs. It may be and has been argued that in the case of unincorporated enterprises, unlike corporations, no significant difference exists between the amount that the entrepreneur withdraws for his consumption or for his individual savings and the amount retained in the business. While this argument does not appear to us theoretically tenable, there is much to be said on practical grounds for measuring only entrepreneurial income and not concerning oneself with entrepreneurial withdrawals as a distinct magnitude. If this is done, entrepreneurial *income* and not entrepreneurial withdrawals enters aggregate income payments to individuals; a new variant of aggregate income payments appears; and net business savings are confined to savings of business corporations. The resulting measures appear in Appendix Table II under the general heading *Adjusted*.

The second group of variants results from the acceptance of the accounting measures of net business savings and the consequent omission of all the adjustments listed and discussed above. If aggregate income payments to individuals include entrepreneurial withdrawals alone, they are not affected by decisions concerning the treatment of net business savings. But if entrepreneurial income, rather than withdrawals, is included, the acceptance of accounting measures of business savings affects also aggregate income payments. For this reason, the entries in Appendix Table II under the general heading *Unadjusted* include not only the new variants of total business savings and hence of national income, but also of aggregate income payments to individuals and of the complementary item, savings of business corporations.

Finally, since business gains and losses on the sale of capital assets are measurable only since 1929, and the adjustment for them affects the continuity of the series, this item is presented separately in Appendix Table III with a distinction between individually owned firms and corporations.

For the various industrial branches distinguished, the number of variants in Appendix Table II

Appendix Table I

NATIONAL INCOME AND AGGREGATE INCOME PAYMENTS TO INDIVIDUALS BY INDUSTRIAL BRANCHES, BASIC VARIANT, 1919-1934

(millions of dollars)

1. Agriculture

	1919	1920	1921	1922	1923	1924	1925	1926	1927	1928	1929	1930	1931	1932	1933	1934
Wages[1]	1,492	1,713	1,097	1,068	1,208	1,189	1,236	1,282	1,295	1,304	1,313	1,112	807	523	484	518
Withdrawals by farm operators	6,648	7,659	5,104	4,882	5,415	5,506	5,559	5,668	5,626	5,614	5,649	5,132	4,084	3,172	3,005	3,356
Interest on mortgages (incl. small amount of dividends)	340	385	391	400	399	399	402	401	412	410	392	373	367	346	324	278
All income payments to individuals	8,480	9,757	6,592	6,350	7,022	7,094	7,197	7,351	7,333	7,328	7,354	6,617	5,258	4,041	3,813	4,152
Net business savings	2,756	-771	-366	-522	-332	231	572	-57	-134	-46	-123	-1,107	-1,770	-1,859	-733	-124
Income originating	11,236	8,986	6,226	5,828	6,690	7,325	7,769	7,294	7,199	7,282	7,231	5,510	3,488	2,182	3,080	4,276

[1] Includes board and perquisites.

2. Mining

	1919	1920	1921	1922	1923	1924	1925	1926	1927	1928	1929	1930	1931	1932	1933	1934
Wages	1,393	1,992	1,448	1,339	1,932	1,616	1,507	1,758	1,546	1,367	1,405	1,173	805	529	533	708
Salaries	156	223	150	161	200	191	197	225	218	201	226	226	183	134	129	169
Employees' compensation	1,548	2,215	1,598	1,500	2,133	1,806	1,704	1,983	1,765	1,568	1,631	1,399	987	663	662	877
Dividends	195	210	193	139	223	213	269	327	278	253	365	249	138	82	75	183
Interest	23	30	38	33	37	51	54	44	44	42	43	43	43	41	39	36
Property income	218	240	231	172	261	264	323	373	322	295	407	292	181	122	114	219
Withdrawals by entrepreneurs	30	39	32	31	34	29	27	26	25	23	21	20	18	14	14	15
All income payments to individuals	1,796	2,494	1,861	1,703	2,428	2,099	2,054	2,382	2,111	1,886	2,059	1,712	1,186	799	790	1,111
Net business savings	36	14	-152	-273	-245	-289	-100	-95	-166	-167	-166	-241	-314	-296	-287	-105
Income originating	1,832	2,508	1,709	1,430	2,183	1,810	1,954	2,288	1,945	1,719	1,893	1,471	872	503	503	1,007

3. Manufacturing

	1919	1920	1921	1922	1923	1924	1925	1926	1927	1928	1929	1930	1931	1932	1933	1934
Wages	9,682	11,587	7,460	7,990	10,160	9,494	9,981	10,318	10,115	10,201	10,899	8,842	6,701	4,636	4,940	6,304
Salaries	2,786	3,114	2,431	2,474	2,862	2,895	2,979	3,229	3,424	3,665	4,013	3,866	3,206	2,424	2,205	2,612
Employees' compensation	12,468	14,701	9,892	10,464	13,022	12,389	12,960	13,547	13,539	13,866	14,912	12,708	9,908	7,060	7,145	8,916
Dividends	1,262	1,489	1,325	1,311	1,764	1,653	1,911	2,119	2,228	2,509	2,578	2,618	1,897	1,117	1,011	1,217
Interest	86	108	138	106	118	155	154	152	154	185	212	237	238	209	193	178
Property income	1,348	1,596	1,464	1,417	1,882	1,807	2,065	2,271	2,383	2,695	2,790	2,856	2,134	1,326	1,203	1,395
Withdrawals by entrepreneurs	475	483	377	375	358	356	338	350	338	333	333	300	246	181	167	198
All income payments to individuals	14,292	16,780	11,732	12,255	15,262	14,553	15,333	16,168	16,260	16,893	18,035	15,864	12,288	8,567	8,516	10,509
Net business savings	1,879	3,126	895	824	1,523	1,045	1,450	1,983	925	1,033	1,793	314	-1,290	-2,320	-1,872	-695
Income originating	16,171	19,907	12,627	13,079	16,784	15,598	16,833	18,151	17,186	17,927	19,828	16,179	10,997	6,248	6,644	9,814

[62]

4. Construction

	1919	1920	1921	1922	1923	1924	1925	1926	1927	1928	1929	1930	1931	1932	1933	1934
Wages and salaries	1,235	1,759	1,288	1,535	2,230	2,243	2,232	2,540	2,456	2,567	2,521	2,159	1,415	643	542	638
Dividends	15	21	32	30	37	32	58	41	47	51	62	83	39	21	19	15
Interest	4.2	5.0	7.8	4.2	5.5	7.3	9.9	11	11	9.5	13	15	13	10	8.7	8.4
Property income	19	26	40	34	43	39	68	52	58	61	74	98	52	31	27	23
Withdrawals by entrepreneurs	287	308	295	406	383	420	552	359	395	435	436	261	176	159	225	276
All income payments to individuals	1,541	2,093	1,623	1,975	2,657	2,702	2,852	2,951	2,909	3,063	3,031	2,518	1,644	833	794	938
Net business savings[1,2]	-40	165	272	-64	-40	189	103	178	220	37	60	128	44	-125	-289	-143
Income originating[2]	1,501	2,258	1,895	1,910	2,617	2,892	2,955	3,128	3,129	3,100	3,091	2,646	1,688	708	506	795

5. Transportation and Other Public Utilities

a. Electric Light and Power and Manufactured Gas

	1919	1920	1921	1922	1923	1924	1925	1926	1927	1928	1929	1930	1931	1932	1933	1934
Wages and salaries	217	261	263	274	346	400	415	464	480	508	533	528	473	379	350	376
Dividends	112	111	112	149	201	241	287	302	339	430	514	605	609	520	431	379
Interest	93	101	112	130	158	194	218	254	278	307	318	340	371	404	407	370
Property income	205	212	224	279	358	435	505	556	617	738	832	944	980	924	839	749
Withdrawals by entrepreneurs	4.0	4.0	3.7	3.2	2.8	2.6	2.3	1.9	1.5	1.4	1.2	1.0	0.8	0.6	0.6	0.5
All income payments to individuals	426	477	491	556	707	837	922	1,022	1,098	1,247	1,367	1,474	1,453	1,304	1,189	1,126
Net business savings[1,2]	-1.3	19	33	64	72	43	148	132	150	160	197	141	49	-43	-43	-35
Income originating[2]	425	496	524	620	778	880	1,070	1,154	1,248	1,407	1,564	1,614	1,503	1,262	1,146	1,091

b. Steam Railroads, Pullman and Express

	1919	1920	1921	1922	1923	1924	1925	1926	1927	1928	1929	1930	1931	1932	1933	1934
Wages, including gratuities[3]	2,462	3,215	2,269	2,116	2,472	2,289	2,324	2,403	2,346	2,264	2,332	1,998	1,584	1,122	1,060	1,165
Salaries[3]	675	865	819	837	870	871	877	893	900	892	895	852	749	564	501	525
Compensation for injuries and pensions[4]	49	73	46	49	59	60	72	77	78	72	74	72	67	59	58	62
Employees' compensation[4]	3,186	4,152	3,135	3,002	3,401	3,220	3,272	3,373	3,324	3,228	3,301	2,922	2,400	1,745	1,619	1,751
Dividends	257	236	209	226	253	277	295	337	437	367	439	435	257	75	76	121
Interest	436	461	477	489	501	525	537	528	528	528	520	535	536	531	478	461
Property income	693	698	685	714	753	802	832	865	965	894	959	970	793	606	554	582
All income payments to individuals[1]	3,879	4,850	3,820	3,716	4,155	4,023	4,104	4,238	4,289	4,122	4,260	3,892	3,193	2,351	2,173	2,333
Net business savings[1]	58	-6.5	-5.2	106	282	243	382	429	176	359	393	-22	-219	-261	-52	-111
Income originating[1]	3,937	4,844	3,814	3,822	4,437	4,265	4,486	4,667	4,465	4,482	4,653	3,870	2,974	2,090	2,122	2,222

c. Other Transportation (Pipe Lines, Street Railways and Water Transportation)

	1919	1920	1921	1922	1923	1924	1925	1926	1927	1928	1929	1930	1931	1932	1933	1934
Wages and salaries	937	1,262	1,048	963	995	1,027	1,007	1,034	1,001	998	1,007	941	805	632	579	628
Dividends	102	78	67	104	93	89	116	103	122	127	143	160	125	133	56	49
Interest	122	121	128	139	145	138	125	114	105	105	104	104	103	104	102	98
Property income	224	199	194	242	239	227	241	217	227	232	247	264	228	237	158	148
Withdrawals by entrepreneurs	9.7	6.7	4.7	4.8	4.7	3.5	4.2	4.3	4.0	4.6	4.6	4.6	3.3	2.3	1.3	2.0
All income payments to Individuals	1,171	1,467	1,247	1,210	1,239	1,257	1,252	1,256	1,231	1,235	1,258	1,209	1,036	872	738	778
Net business savings[1]	44	70	3.0	3.3	35	33	33	38	14	31	42	-66	-75	-127	-31	-37
Income originating[1]	1,215	1,538	1,250	1,213	1,274	1,291	1,285	1,293	1,245	1,266	1,299	1,143	961	745	707	740

d. Communication

	1919	1920	1921	1922	1923	1924	1925	1926	1927	1928	1929	1930	1931	1932	1933	1934
Wages and salaries	307	404	402	426	471	504	527	567	586	625	702	710	635	529	455	477
Pensions and benefits	4.1	5.4	5.4	6.0	6.1	6.0	6.6	7.2	7.6	8.3	9.3	9.6	11	10	9.9	11
Employees' compensation	311	409	408	432	477	510	534	574	593	634	711	719	646	540	465	489
Dividends	51	51	58	70	81	92	103	112	125	130	146	171	188	191	188	187
Interest	29	33	36	29	32	33	41	41	42	37	36	34	37	48	50	51
Property income	80	84	94	99	113	125	144	153	167	167	181	204	225	239	238	238
All income payments to Individuals	391	493	502	532	590	634	677	728	760	801	892	924	871	779	703	727
Net business savings[1]	26	21	30	47	47	43	69	76	81	102	109	46	12	-54	-67	-59
Income originating[1]	417	514	532	578	636	677	746	804	841	902	1,001	970	883	724	636	667

e. Total

	1919	1920	1921	1922	1923	1924	1925	1926	1927	1928	1929	1930	1931	1932	1933	1934
All income payments to Individuals	5,867	7,287	6,059	6,014	6,690	6,751	6,956	7,243	7,378	7,405	7,777	7,498	6,554	5,306	4,803	4,964
Net business savings	122	174	311	214	399	376	663	691	460	638	736	217	-159	-445	-220	-310
Income originating	5,989	7,461	6,371	6,228	7,089	7,127	7,624	7,934	7,838	8,043	8,513	7,715	6,395	4,860	4,583	4,654

[1] Not adjusted for gains and losses on inventory holdings.
[2] The figures for the electric light and power industry are based on Census data and are for operating companies alone. Their profits and losses from the sale of capital assets cannot be estimated.
[3] Owing to the reclassification of railroad employees, the 1932-34 figures are not strictly comparable with those for the earlier years. Comparable 1929-31 figures, in millions of dollars are:

	1929	1930	1931
Wages	2,347	2,011	1,596
Salaries	880	838	737

[4] Including compensation for injuries to persons other than employees.

6. Trade

	1919	1920	1921	1922	1923	1924	1925	1926	1927	1928	1929	1930	1931	1932	1933	1934
Wages and salaries	5,403	6,021	5,138	5,641	6,337	6,471	6,916	7,341	7,171	7,359	7,797	7,431	6,448	4,967	4,363	4,822
Dividends	401	382	320	302	369	390	440	471	495	499	566	497	386	214	179	214
Interest	33	42	40	39	22	31	31	25	31	43	56	59	66	61	49	45
Property income	434	424	360	341	391	420	471	496	526	542	621	556	452	275	228	260
Withdrawals by entrepreneurs	2,197	2,595	2,177	2,050	2,073	2,138	2,091	2,134	2,109	2,128	2,232	2,191	2,026	1,774	1,648	1,667
All income payments to individuals	8,034	9,039	7,676	8,032	8,802	9,029	9,478	9,970	9,807	10,030	10,650	10,178	8,926	7,016	6,239	6,749
Net business savings[1]	2,081	2,548	2,017	530	1,290	713	605	1,516	658	866	527	723	-137	-1,010	-1,393	-930
Income originating	10,115	11,587	9,692	8,562	10,091	9,742	10,082	11,486	10,465	10,895	11,177	10,901	8,789	6,006	4,846	5,819

7. Finance

a. Banking

	1919	1920	1921	1922	1923	1924	1925	1926	1927	1928	1929	1930	1931	1932	1933	1934
Wages and salaries	362	452	495	492	520	550	573	607	642	673	698	679	605	516	453	455
Dividends paid	265	286	300	315	312	313	329	342	370	381	467	450	411	280	158	177
All income payments to individuals	627	738	795	808	832	863	903	948	1,012	1,054	1,165	1,128	1,016	796	611	632
Net business savings[1]	224	200	107	85	106	139	197	197	180	237	151	-2.5	-231	-280	-204	-160
Income originating[1]	852	938	902	893	938	1,002	1,100	1,145	1,192	1,290	1,316	1,126	786	516	407	472

b. Insurance

	1919	1920	1921	1922	1923	1924	1925	1926	1927	1928	1929	1930	1931	1932	1933	1934
Wages and salaries	540	656	680	698	802	898	995	1,078	1,143	1,210	1,286	1,284	1,194	1,014	921	969
Dividends[2]	21	22	26	49	44	37	43	48	56	62	70	61	61	31	16	28
Interest	-51	-63	-70	-73	-74	-82	-83	-81	-85	-90	-83	-82	-88	-82	-78	-87
Property income	-30	-42	-45	-24	-31	-45	-40	-33	-29	-28	-13	-21	-27	-51	-62	-59
All income payments to individuals	510	615	635	674	771	854	955	1,045	1,114	1,182	1,273	1,263	1,167	963	859	911
Net business savings[1]	46	-62	-85	-93	-103	-106	-48	-45	36	101	20	-85	-100	-75	-29	-30
Income originating[1]	556	553	550	581	668	748	906	1,000	1,150	1,283	1,293	1,178	1,066	888	830	881

c. Real Estate

	1919	1920	1921	1922	1923	1924	1925	1926	1927	1928	1929	1930	1931	1932	1933	1934
Wages and salaries	611	652	617	599	651	698	642	752	862	981	1,115	896	753	568	501	569
Dividends	52	103	89	124	202	182	218	211	202	206	238	154	122	59	30	23
Interest on corporate long term debt	298	288	279	246	280	273	296	300	315	426	433	435	451	375	252	228
Interest on individuals' mortgages	423	481	513	601	738	900	1,072	1,216	1,447	1,571	1,711	1,727	1,683	1,573	1,206	1,001
Property income	774	871	880	972	1,221	1,355	1,587	1,728	1,964	2,204	2,382	2,315	2,256	2,007	1,488	1,252
Net rentals received by individuals	1,804	1,912	1,978	2,271	2,360	2,469	2,471	2,326	2,176	2,241	2,216	1,733	1,136	705	754	706
Net imputed rent received by individuals	866	1,087	1,318	1,483	1,555	1,700	1,676	1,576	1,499	1,547	1,547	1,378	953	614	557	459
Entrepreneurial income	2,670	2,999	3,296	3,754	3,915	4,169	4,146	3,902	3,675	3,788	3,763	3,111	2,088	1,319	1,311	1,165
All income payments to individuals	4,055	4,522	4,794	5,324	5,786	6,223	6,374	6,382	6,500	6,973	7,260	6,323	5,097	3,894	3,301	2,985
Net business savings[1]	14	-41	-35	15	-45	17	-98	-60	-57	0.6	-329	-286	-352	-410	-353	-140
Income originating[1]	4,069	4,480	4,759	5,339	5,742	6,240	6,276	6,322	6,443	6,974	6,932	6,037	4,746	3,484	2,947	2,845

d. Total

	1919	1920	1921	1922	1923	1924	1925	1926	1927	1928	1929	1930	1931	1932	1933	1934
All income payments to individuals	5,193	5,874	6,223	6,806	7,390	7,939	8,232	8,375	8,626	9,208	9,698	8,714	7,280	5,653	4,771	4,528
Net business savings	205	44	75	4.0	-26	58	72	114	160	343	-157	-371	-636	-747	-600	-335
Income originating	5,397	5,918	6,299	6,810	7,363	7,997	8,304	8,489	8,786	9,551	9,541	8,344	6,644	4,906	4,171	4,193

[1] Not adjusted for gains and losses on inventory holdings.
[2] Excluding payments to policy holders.

8. Government

	1919	1920	1921	1922	1923	1924	1925	1926	1927	1928	1929	1930	1931	1932	1933	1934
Wages and salaries, Federal	2,071	1,534	1,323	1,142	1,149	1,179	1,241	1,289	1,308	1,348	1,402	1,430	1,448	1,364	1,220	1,408
Wages and salaries, state	155	180	193	195	206	223	247	238	267	289	308	320	326	326	317	311
Wages and salaries, county	149	177	209	223	232	247	243	259	281	304	326	341	358	351	319	328
Wages and salaries, city, township and minor civil divisions	583	684	727	740	792	846	887	952	1,028	1,073	1,128	1,148	1,151	1,121	995	1,040
Wages and salaries, public education	695	825	998	1,134	1,196	1,264	1,350	1,444	1,530	1,605	1,666	1,713	1,725	1,665	1,512	1,464
Pensions and relief[1]	422	558	668	675	690	713	644	633	663	687	726	769	875	907	1,852	2,691
Employees' compensation	4,076	3,958	4,119	4,109	4,265	4,473	4,617	4,815	5,077	5,306	5,555	5,721	5,883	5,734	6,215	7,242
Interest	1,126	1,340	1,373	1,444	1,463	1,423	1,441	1,452	1,437	1,435	1,472	1,488	1,458	1,531	1,623	1,732
All income payments to individuals	5,202	5,299	5,492	5,553	5,728	5,896	6,058	6,267	6,514	6,741	7,028	7,209	7,341	7,265	7,838	8,974
Net savings	-4,279	1,607	660	1,002	1,425	1,465	1,571	1,877	2,017	1,764	1,507	975	-700	-1,406	-818	-1,181
Income originating	923	6,906	6,151	6,555	7,153	7,361	7,628	8,144	8,531	8,505	8,535	8,184	6,641	5,859	7,020	7,793

[1] Relief payments included are as follows, in millions of dollars:

	1933	1934
Work relief	619	1,389
Direct relief	482	657

9. Service

	1919	1920	1921	1922	1923	1924	1925	1926	1927	1928	1929	1930	1931	1932	1933	1934
Wages, salaries and withdrawals by entrepreneurs	5,157	6,207	5,966	7,178	7,406	8,086	8,820	9,088	9,325	9,925	10,547	9,763	8,924	7,641	7,986	7,996
Dividends	29	80	66	49	68	71	92	105	107	99	117	113	74	54	34	49
Interest	16	15	18	23	28	33	41	48	63	73	87	99	94	110	95	85
Property Income	45	95	84	72	96	104	133	153	169	172	204	212	167	165	129	134

9. Service (Continued)

	1919	1920	1921	1922	1923	1924	1925	1926	1927	1928	1929	1930	1931	1932	1933	1934
All income payments to individuals	5,202	6,302	6,050	7,250	7,502	8,191	8,953	9,241	9,495	10,097	10,751	9,975	9,092	7,806	8,115	8,130
Net business savings	1,211	720	358	442	788	607	658	954	520	597	473	162	-490	-1,430	-2,005	-895
Income originating	6,413	7,022	6,408	7,691	8,290	8,798	9,611	10,195	10,015	10,693	11,224	10,136	8,602	6,375	6,111	7,236

10. Miscellaneous

	1919	1920	1921	1922	1923	1924	1925	1926	1927	1928	1929	1930	1931	1932	1933	1934
Wages, salaries and withdrawals by entrepreneurs	1,753	1,968	1,681	1,900	2,145	2,238	2,451	2,603	2,646	2,838	3,074	2,919	2,646	2,245	2,067	2,302
Dividends	50	56	51	52	66	65	76	82	87	93	103	112	-37	-98	-125	-146
Interest	61	63	67	65	70	75	79	80	83	93	97	101	104	101	89	82
Dividends and interest, international	29	43	69	86	94	129	154	110	132	149	151	202	284	252	169	93
Property income	140	162	187	203	230	270	309	272	302	335	351	415	350	254	133	29
All income payments to individuals	1,893	2,130	1,868	2,103	2,375	2,508	2,759	2,874	2,948	3,173	3,425	3,334	2,996	2,499	2,200	2,330
Net business savings	371	285	147	56	164	99	102	208	86	84	-347	-1,123	-1,113	-962	-839	-82
Income originating	2,264	2,416	2,014	2,159	2,538	2,607	2,861	3,083	3,034	3,257	3,077	2,211	1,883	1,537	1,361	2,248

11. Total National Income

	1919	1920	1921	1922	1923	1924	1925	1926	1927	1928	1929	1930	1931	1932	1933	1934
Aggregate income payments to individuals	57,499	67,056	55,177	58,041	65,854	66,763	69,921	72,823	73,381	75,823	79,808	73,620	62,565	49,785	47,880	52,385
Net savings of enterprises (aggregate of estimates by industrial branches)	4,343	7,913	4,216	2,213	4,946	4,494	5,701	7,369	4,747	5,147	4,303	-323	-6,566	-10,601	-9,055	-4,551
Net savings of enterprises, adjusted for disparity between depreciation and depletion at book values and at reproduction prices	2,427	5,330	3,166	1,665	3,853	3,606	4,926	6,654	4,048	4,574	3,616	-680	-6,556	-10,157	-8,596	-4,536
National income	59,926	72,386	58,343	59,706	69,706	70,369	74,846	79,477	77,429	80,397	83,424	72,940	56,010	39,628	39,283	47,849

Appendix Table II

NATIONAL INCOME AND AGGREGATE INCOME PAYMENTS TO INDIVIDUALS, OTHER VARIANTS, 1919-1934

(millions of dollars)

1. Agriculture

Adjusted

	1919	1920	1921	1922	1923	1924	1925	1926	1927	1928	1929	1930	1931	1932	1933	1934
Entrepreneurial income[1]	9,404	6,888	4,738	4,360	5,083	5,737	6,131	5,611	5,492	5,568	5,526	4,025	2,314	1,313	2,272	3,480
All income payments to individuals, second variant	11,236	8,986	6,226	5,828	6,690	7,325	7,769	7,294	7,199	7,282	7,231	5,510	3,488	2,182	3,080	4,276

[1] Including a small amount of corporate savings which cannot be segregated.

2. Mining

Adjusted

	1919	1920	1921	1922	1923	1924	1925	1926	1927	1928	1929	1930	1931	1932	1933	1934
Entrepreneurial income	50	61	37	27	33	27	40	43	34	31	35	24	15	9.0	12	13
All income payments to individuals, second variant	1,815	2,516	1,866	1,700	2,426	2,097	2,067	2,400	2,120	1,894	2,073	1,716	1,183	794	788	1,109
Net savings of corporations	17	-7.5	-157	-269	-243	-287	-113	-112	-175	-175	-180	-245	-312	-290	-284	-103

Unadjusted

	1919	1920	1921	1922	1923	1924	1925	1926	1927	1928	1929	1930	1931	1932	1933	1934
Net business savings, second variant	4.3	179	-462	-150	-296	-298	-54	-80	-269	-145	-152	-303	-399	-312	-255	-55
Income originating, second variant	1,800	2,673	1,399	1,553	2,132	1,801	2,000	2,303	1,842	1,741	1,907	1,409	787	487	535	1,056
Entrepreneurial income, second variant	48	71	19	34	30	26	42	44	29	32	35	22	11	7.7	13	15
All income payments to individuals, third variant	1,813	2,526	1,848	1,707	2,423	2,097	2,069	2,401	2,115	1,895	2,073	1,714	1,180	793	789	1,111
Net savings of corporations, second variant	-13	148	-449	-153	-291	-296	-69	-98	-273	-154	-166	-305	-393	-306	-254	-55

3. Manufacturing

Adjusted

	1919	1920	1921	1922	1923	1924	1925	1926	1927	1928	1929	1930	1931	1932	1933	1934
Entrepreneurial income	833	845	562	513	585	482	521	534	475	439	491	366	163	4.8	17	83
All income payments to individuals, second variant	14,650	17,142	11,917	12,394	15,489	14,678	15,546	16,352	16,398	17,000	18,192	15,930	12,205	8,391	8,366	10,394
Net savings of corporations	1,521	2,765	710	685	1,295	920	1,286	1,799	788	927	1,635	249	-1,208	-2,144	-1,722	-580

Unadjusted

	1919	1920	1921	1922	1923	1924	1925	1926	1927	1928	1929	1930	1931	1932	1933	1934
Net business savings, second variant	2,781	1,026	-1,918	-1,196	1,652	869	1,501	1,201	542	1,058	1,473	-1,858	-3,029	-3,182	-1,021	393
Income originating, second variant	17,073	17,807	9,814	13,451	16,913	15,422	16,884	17,369	16,803	17,952	19,507	14,006	9,258	5,385	7,495	10,901
Entrepreneurial income, second variant	951	596	220	557	599	464	526	461	439	441	431	183	30	-55	82	127
All income payments to individuals, third variant	14,768	16,893	11,575	12,438	15,503	14,660	15,551	16,279	16,362	17,002	18,133	15,747	12,072	8,331	8,430	10,437
Net savings of corporations, second variant	2,305	914	-1,761	1,013	1,410	762	1,332	1,090	441	950	1,374	-1,741	-2,813	-2,946	-936	464

4. Construction

Adjusted

	1919	1920	1921	1922	1923	1924	1925	1926	1927	1928	1929	1930	1931	1932	1933	1934
Entrepreneurial income	253	439	539	348	337	551	616	470	538	444	468	370	244	141	27	164
All income payments to individuals, second variant	1,507	2,224	1,868	1,917	2,610	2,834	2,916	3,062	3,052	3,072	3,063	2,627	1,712	815	597	826
Net savings of corporations	-6.0	34	27	-6.2	6.4	58	39	66	77	28	29	18	-24	-108	-91	-32

Unadjusted

	1919	1920	1921	1922	1923	1924	1925	1926	1927	1928	1929	1930	1931	1932	1933	1934
Net business savings, second variant	116	56	-42	36	67	140	130	115	97	84	77	-37	-99	-204	-172	-44
Income originating, second variant	1,637	2,149	1,581	2,010	2,724	2,843	2,982	3,065	3,006	3,147	3,108	2,481	1,545	629	623	893
Entrepreneurial income, second variant	381	352	280	433	427	510	639	418	440	482	478	251	147	84	138	253
All income payments to individuals, third variant	1,635	2,137	1,609	2,002	2,700	2,793	2,939	3,010	2,954	3,110	3,072	2,508	1,615	757	707	915
Net savings of corporations, second variant	22	12	-28	8.8	23	50	43	55	52	37	35	-27	-70	-129	-84	-22

5. Transportation and Other Public Utilities

a. Other Transportation (Pipe Lines, Street Railways and Water Transportation)

Adjusted[1]

	1919	1920	1921	1922	1923	1924	1925	1926	1927	1928	1929	1930	1931	1932	1933	1934
Entrepreneurial income	14	10	2.5	3.8	4.2	3.7	4.8	5.1	3.3	4.9	5.9	2.6	1.3	0.03	2.0	1.6
All income payments to individuals, second variant	1,175	1,471	1,245	1,209	1,238	1,258	1,253	1,256	1,231	1,235	1,259	1,207	1,034	870	739	777
Net savings of corporations	40	67	5.2	4.3	36	33	33	37	15	31	40	-64	-73	-125	-31	-37

Unadjusted

	1919	1920	1921	1922	1923	1924	1925	1926	1927	1928	1929	1930	1931	1932	1933	1934
Net business savings, second variant	44	70	3.0	3.3	35	33	33	38	14	31	43	-68	-84	-130	-40	-50
Income originating, second variant	1,215	1,538	1,250	1,213	1,274	1,291	1,285	1,293	1,245	1,266	1,301	1,141	952	742	699	728
Entrepreneurial income, second variant	14	10	2.5	3.8	4.2	3.7	4.8	5.1	3.3	4.9	6.0	2.5	0.5	-0.2	1.2	0.9
All income payments to individuals, third variant	1,175	1,471	1,245	1,209	1,238	1,258	1,253	1,256	1,231	1,235	1,259	1,207	1,033	869	738	777
Net savings of corporations, second variant	40	67	5.2	4.3	36	33	33	37	15	31	41	-66	-81	-127	-39	-49

b. Total

Adjusted

	1919	1920	1921	1922	1923	1924	1925	1926	1927	1928	1929	1930	1931	1932	1933	1934
Entrepreneurial income[1]	18	14	6.2	7.0	7.0	6.3	7.1	7.0	4.8	6.2	7.1	3.7	2.1	0.7	2.5	2.1
All income payments to individuals, second variant	5,871	7,291	6,057	6,013	6,690	6,752	6,956	7,244	7,377	7,405	7,778	7,496	6,552	5,304	4,804	4,963
Net savings of corporations	118	171	314	215	400	375	668	690	461	637	735	219	-156	-443	-220	-309

Unadjusted

	1919	1920	1921	1922	1923	1924	1925	1926	1927	1928	1929	1930	1931	1932	1933	1934
Net business savings, second variant	127	104	61	220	436	362	632	674	421	653	741	97	-244	-490	-209	-261
Income originating, second variant	5,994	7,391	6,121	6,234	7,126	7,113	7,538	7,917	7,799	8,058	8,518	7,596	6,309	4,816	4,595	4,703
Entrepreneurial income, second variant	18	14	6.2	7.0	7.0	6.3	7.1	7.0	4.8	6.2	7.2	3.5	1.3	0.4	1.8	1.5
All income payments to individuals, third variant	5,871	7,291	6,057	6,013	6,690	6,752	6,956	7,244	7,377	7,405	7,778	7,496	6,551	5,303	4,803	4,963
Net savings of corporations, second variant	123	101	64	221	437	361	632	673	422	652	740	99	-241	-487	-208	-260

[1] Not adjusted for gains and losses on inventory holdings.

6. Trade

Adjusted

	1919	1920	1921	1922	1923	1924	1925	1926	1927	1928	1929	1930	1931	1932	1933	1934
Entrepreneurial income	3,963	4,231	3,577	2,405	2,907	2,578	2,439	3,006	2,504	2,575	2,560	2,637	2,110	1,381	1,020	1,223
All income payments to individuals, second variant	9,800	10,675	9,076	8,387	9,635	9,469	9,826	10,843	10,201	10,477	10,979	10,624	9,011	6,623	5,611	6,305
Net savings of corporations	315	912	616	175	456	273	257	643	264	418	198	277	-221	-617	-765	-486

Unadjusted

	1919	1920	1921	1922	1923	1924	1925	1926	1927	1928	1929	1930	1931	1932	1933	1934
Net business savings, second variant	3,050	428	-920	890	1,301	798	934	605	464	719	241	-1,182	-1,775	-1,870	-414	-66
Income originating, second variant	11,084	9,467	6,755	8,922	10,102	9,827	10,411	10,575	10,271	10,748	10,891	8,996	7,152	5,147	5,825	6,682
Entrepreneurial income, second variant	4,623	3,014	1,718	2,633	2,914	2,630	2,631	2,502	2,400	2,498	2,404	1,639	1,236	942	1,501	1,643
All income payments to individuals, third variant	10,460	9,458	7,217	8,615	9,642	9,521	10,018	10,339	10,097	10,400	10,822	9,626	8,137	6,184	6,091	6,725
Net savings of corporations, second variant	624	8.5	-462	307	460	306	394	236	174	348	69	-630	-985	-1,037	-267	-43

7. Finance

Unadjusted

	1919	1920	1921	1922	1923	1924	1925	1926	1927	1928	1929	1930	1931	1932	1933	1934
Net business savings, second variant	285	97	-13	7.0	-42	50	50	92	159	339	67	-350	-940	-1,159	-958	-604
Income originating, second variant	5,477	5,971	6,211	6,813	7,347	7,989	8,282	8,467	8,785	9,547	9,765	8,365	6,340	4,495	3,813	3,924

8. Service

Adjusted

	1919	1920	1921	1922	1923	1924	1925	1926	1927	1928	1929	1930	1931	1932	1933	1934
Wages, salaries and entrepreneurial income	6,333	6,926	6,328	7,595	8,153	8,646	9,418	10,013	9,850	10,508	11,028	9,992	8,570	6,583	6,299	7,130
All income payments to individuals, second variant	6,377	7,021	6,412	7,666	8,249	8,751	9,551	10,166	10,020	10,680	11,232	10,203	8,737	6,747	6,428	7,264
Net savings of corporations	35	0.6	-4.0	25	41	47	60	29	-4.4	13	-7.9	-67	-136	-372	-317	-28

Unadjusted

	1919	1920	1921	1922	1923	1924	1925	1926	1927	1928	1929	1930	1931	1932	1933	1934
Net business savings, second variant	1,225	725	337	443	790	604	653	948	518	591	496	149	-561	-1,535	-2,045	-884
Income originating, second variant	6,427	7,027	6,387	7,692	8,292	8,795	9,606	10,189	10,013	10,687	11,247	10,124	8,531	6,271	6,071	7,246
Net savings of corporations, second variant	49	5.6	-25	26	43	44	55	23	-6.4	7.0	15	-80	-207	-477	-357	-18

9. Miscellaneous

Adjusted

	1919	1920	1921	1922	1923	1924	1925	1926	1927	1928	1929	1930	1931	1932	1933	1934
Wages, salaries and entrepreneurial income	2,060	2,204	1,862	1,933	2,266	2,312	2,507	2,768	2,713	2,896	3,112	2,858	2,495	2,025	2,003	2,335
All income payments to individuals, second variant	2,200	2,366	2,048	2,136	2,496	2,582	2,816	3,040	3,015	3,231	3,463	3,273	2,845	2,279	2,136	2,364
Net savings of corporations	65	50	-34	23	42	26	45	43	20	26	-385	-1,062	-962	-742	-775	-116

Unadjusted

	1919	1920	1921	1922	1923	1924	1925	1926	1927	1928	1929	1930	1931	1932	1933	1934
Net business savings, second variant	375	261	127	62	163	99	105	207	86	84	81	-1,376	-1,866	-1,725	-1,491	-216
Income originating, second variant	2,268	2,392	1,994	2,165	2,537	2,607	2,865	3,082	3,034	3,257	3,506	1,958	1,130	774	708	2,114
Net savings of corporations, second variant	69	26	-54	29	41	26	49	42	20	26	43	-1,315	-1,715	-1,505	-1,427	-250

10. Total National Income

Adjusted

	1919	1920	1921	1922	1923	1924	1925	1926	1927	1928	1929	1930	1931	1932	1933	1934
Entrepreneurial income[1] (aggregate of estimates by industrial branches)	17,191	15,476	12,756	11,414	12,867	13,550	13,900	13,574	12,723	12,852	12,850	10,537	6,936	4,169	4,662	6,131
Entrepreneurial income[1], adjusted for disparity between depreciation and depletion at book value and at reproduction prices	16,275	14,283	12,276	11,156	12,414	13,152	13,515	13,259	12,414	12,569	12,543	10,370	6,866	4,163	4,731	6,044

[1] Excluding entrepreneurial income in service and miscellaneous where the item of withdrawals cannot be segregated from wages and salaries.

10. Total National Income
Adjusted (Continued)

	1919	1920	1921	1922	1923	1924	1925	1926	1927	1928	1929	1930	1931	1932	1933	1934
Aggregate income payments to individuals[2], second variant (aggregate of estimates by industrial branches)	63,852	69,393	57,186	58,400	67,403	68,322	71,736	75,042	74,522	76,990	80,737	73,304	60,354	46,054	44,417	51,003
Aggregate income payments to individuals[2], second variant, adjusted for disparity between depreciation and depletion at book value and at reproduction prices	62,936	68,200	56,706	58,142	66,950	67,924	71,351	74,727	74,213	76,707	80,430	73,137	60,284	46,048	44,486	50,916
Net savings of corporations and government (aggregate of estimates by industrial branches)	-2,010	5,576	2,207	1,854	3,396	2,935	3,885	5,150	3,607	3,980	3,374	-6.6	-4,355	-6,870	-5,593	-3,169
Net savings of corporations and government, adjusted for disparity between depreciation and depletion at book value and at reproduction prices	-3,010	4,186	1,637	1,564	2,756	2,445	3,495	4,750	3,217	3,690	2,994	-197	-4,275	-6,420	-5,203	-3,067

Unadjusted

	1919	1920	1921	1922	1923	1924	1925	1926	1927	1928	1929	1930	1931	1932	1933	1934
Net savings of enterprises, second variant	6,441	3,713	-2,537	3,184	5,164	4,320	6,095	5,582	3,902	5,099	4,407	-4,991	-11,383	-13,742	-8,115	-2,796
Income originating, second variant	63,940	70,769	52,640	61,225	71,017	71,083	76,015	78,405	77,283	80,922	84,215	68,629	51,182	36,044	39,764	49,588
Entrepreneurial income[1], second variant	18,095	13,933	10,278	11,778	12,975	13,542	14,122	12,946	12,480	12,816	12,644	9,235	5,827	3,610	5,318	6,684
Aggregate income payments to individuals[2], third variant	64,756	67,850	54,708	58,764	67,511	68,315	71,958	74,414	74,279	76,954	80,531	72,001	59,246	45,496	45,073	51,557
Net savings of corporations and government, second variant	-816	2,919	-2,068	2,461	3,506	2,769	4,057	3,991	3,005	3,968	3,684	-3,372	-8,063	-9,452	-5,308	-1,968

[1] Excluding entrepreneurial income in service and miscellaneous where the item of withdrawals cannot be segregated from wages and salaries.
[2] Including entrepreneurial income in service and miscellaneous.

and the number of entries in Appendix Table III differ. A few brief explanatory comments will perhaps be of assistance in consulting the tables.

For *agriculture* Appendix Table II provides only one variant, showing entrepreneurial income and aggregate income payments, both of which include entrepreneurial business savings. No segregation of the small amount of corporate savings is possible for agriculture; and since the estimates of agricultural income are based upon a direct comparison of gross income with expenses, no adjustments for gains and losses on inventory holdings or on the sale of capital assets are necessary. Hence there is no entry for the latter item in Appendix Table III. The adjustment for the disparity between depreciation and depletion at book value and at reproduction prices cannot be made for the separate industrial divisions, a point to be remembered for all the specific industrial branches commented upon below.

For *mining, manufacturing* and *construction*, three industrial branches in which the activity of individual entrepreneurs can be measured, Appendix Table II provides the full list of variants: the new variants of aggregate income payments inclusive of business savings by individual entrepreneurs, adjusted for gains and losses on inventory holdings and on the sale of capital assets (the latter since 1929); the first variant of corporate savings, similarly adjusted; and the unadjusted measures of income originating, net business savings, net corporate savings, and aggregate income payments, including unadjusted business savings of individual entrepreneurs. For these three industrial branches full entries are made in Appendix Table III.

For the various divisions of the *transportation and other public utilities* group no adjustment of business savings for gains and losses on inventory holdings was possible; consequently the unadjusted variants in Appendix Table II are fewer. In the *electric light and power and manufactured gas* division, withdrawals and total income of the few individual entrepreneurs were treated as identical; hence no adjusted variants of entrepreneurial income and of aggregate income payments appear for this division in Appendix Table II. The only adjustment made in net business savings, as shown in Appendix Table I, was for profits and losses on the sale of capital assets in the manufactured gas industry (see the entry in Appendix Table III); and it therefore seemed unnecessary to show any unadjusted variants for this division in Appendix Table II. In *steam railroads* (including *Pullman and express*) and *communication,* there are no individual entrepreneurs; and the estimates of business savings, based on the data of the Interstate Commerce Commission and the Census of Electrical Industries, already exclude profits and losses on the sale of capital assets throughout the period. Since in these divisions the specific adjustment of business savings for gains and losses on inventory holdings is impossible, no variants appear for these divisions in Appendix Table II. In the *other transportation* division alone, where in *water transportation* entrepreneurial income is considerable and profits and losses on the sale of capital assets can be segregated since 1929, Appendix Table II does provide a full list of variants. Complete data are provided for the *total transportation and other public utilities* group in Appendix Tables II and III.

For *trade* a full list of variants is shown in Appendix Table II and a full list of entries in Appendix Table III (see comments above for mining, manufacturing and construction).

For the separate divisions of *finance* no adjustment of business savings for gains and losses on inventory holdings was possible. Also, the activity of individual entrepreneurs was not covered in *banking* and *insurance* (but insurance agents were included); in *real estate* income payments to individual entrepreneurs alone (rather than total entrepreneurial income) were estimated. Hence, the only adjustment possible for the separate divisions of finance was that for profits and losses on the sale of capital assets since 1929. As the corresponding items are given in Appendix Table III, it was deemed unnecessary to provide unadjusted variants in Appendix Table II. The only variant in the latter table is for the whole finance group, an unadjusted variant of business savings and of income originating. These variants differ from the corresponding items in Appendix Table I in that they are not adjusted for gains and losses either on inventory holdings or on the sale of capital assets.

The measures of income originating in the field of *government* are not subject to the distortions affecting business savings; and no distinction between entrepreneurial withdrawals and savings is here involved. Hence, no variants appear for this field in Appendix Table II and no entries in Appendix Table III.

For the *service* and *miscellaneous* groups rough measures of net business savings of individual entrepreneurs were obtained, although, like most of the estimates in these two fields, they rest upon very slender foundations. But since in these fields

APPENDIX

Appendix Table III

ESTIMATED LOSSES AND GAINS BY BUSINESS ENTERPRISES ON THE SALE OF CAPITAL ASSETS, 1929-1934

(millions of dollars)

	1929	1930	1931	1932	1933	1934
Mining						
Corporate	31	32	−12	−18	−14	−3.2
Individual entrepreneurs	1.5	1.8	−0.6	−1.4	−0.9	−0.2
Manufacturing						
Corporate	71	−56	−204	−218	−236	117
Individual entrepreneurs	−30	−19	−18	−12	−24	−38
Construction						
Corporate	4.6	−9.3	−19	−12	−8.1	−2.1
Individual entrepreneurs	3.8	−4.1	−8.1	−6.8	−8.6	−2.2
Transportation and other public utilities						
a Manufactured gas						
Corporate	−.008	0.3	−2.9	−1.8	−7.0	−5.2
b Water transportation						
Corporate	1.2	−1.5	−8.0	−2.3	−7.7	−12
Individual entrepreneurs	0.1	−0.2	−0.8	−0.2	−0.8	−0.6
c Total						
Corporate	1.2	−1.3	−11	−4.1	−15	−17
Individual entrepreneurs	0.1	−0.2	−0.8	−0.2	−0.8	−0.6
Trade						
Corporate	28	−29	−100	−55	−62	−9.9
Individual entrepreneurs	29	−28	−86	−47	−49	−7.8
Finance						
a Banking						
Corporate	48	18	−83	−140	−142	−112
b Insurance						
Corporate	35	−14	−70	−109	−114	−117
c Real estate						
Corporate	141	19	−105	−144	−115	−46
d Total						
Corporate	224	24	−257	−393	−371	−274
Service						
Corporate	23	2.4	−42	−91	−56	−2.8
Miscellaneous						
Corporate	428	−251	−751	−762	−654	−134
Total						
Corporate	811	−287	−1,396	−1,553	−1,417	−327
Individual entrepreneurs	5.0	−49	−114	−67	−83	−49

corporations account for only a minor share of activity, it was deemed inadvisable to use the corporate data for the purpose of adjusting business savings of individual entrepreneurs for the gains and losses on inventory holdings and on the sale of capital assets. Instead, adjusted and unadjusted business savings of individual entrepreneurs were treated as identical and entered under the heading of *Adjusted* in Appendix Table II. This treatment explains why, for these two fields, Appendix Table II does not provide the full list of variants; it omits the third variant of aggregate income payments (which in these fields is identical with the second variant) and the second variant of net savings of corporations (which is identical with the first). For the same reason the entries in Appendix Table III are confined to corporations in these two fields.

For the national totals the adjusted variants for the separate industrial branches can be further corrected for the disparity between depreciation charges at book value and at reproduction cost. The unadjusted variants are, for the national totals, the same as those for the various industrial branches.

It is hoped that the brief comments above and the detailed presentation in Appendix Tables I, II and III will enable the student to distinguish clearly each adjustment and to recombine the sub-component elements so as to arrive at other variants of national income that would better satisfy his purpose.

The detailed and basic analysis was carried only through 1934. The measures for 1935, which appear in Tables 1, 3, 4 and 6, were obtained by carrying forward our estimates for 1934 on the basis of the changes from 1934 to 1935 shown in the corresponding estimates of the Department of Commerce. This extrapolation was based on the Department's most recent estimates for 1934 and 1935.[1]

[1] See *National Income, 1929–1936*, U. S. Department of Commerce (Washington, 1937).

APPENDIX B
COMPARISON WITH DEPARTMENT OF COMMERCE ESTIMATES

COMPARISONS of some of our estimates with the measures of national income for 1929–35 published by the Department of Commerce,[1] when made not only for the over-all totals but also for the various industrial branches or types of income share, reveal discrepancies that stem from three essentially distinct sources: differences in (1) scope and concept; (2) the estimating procedures and the data used; (3) the industrial or type-of-payment classification, especially the former.

To enumerate the various sources of discrepancies between the two sets of estimates, and to demonstrate the quantitative effect of each source would be impossible here. Especially would analysis of differences in method, data and classification be a task much beyond the scope of this report, for it would require a careful and detailed description of sources and methods used in deriving both sets of estimates, and a minute comparison of the two. The only observation that can be made here is that our attempt to cover the entire period 1919–35 was rendered difficult by the lack of a number of sources for years before 1929 that are available for the period since 1929; and that in order to arrive at comparable and continuous series for the entire period the methods employed for years preceding and following 1929 had to be in close consonance. For this reason alone the data and methods used in our study naturally differ from those used in a study that confines itself to years since 1929. These differences in data and method entail also different classifications. But it is feasible here to indicate the basic differences in scope and concept between our estimates and those of the Department of Commerce, and to show how they are reflected in quantitative differences.

The basic differences in scope between the magnitude that is here designated as aggregate income payments to individuals and that designated by the Department of Commerce (and in our earlier writings) as income paid out are as follows: (1) aggregate income payments to individuals include imputed rental on houses inhabited by owners, while income paid out omits it; (2) our aggregate includes all relief payments by governmental agencies, not merely work-relief, which is the only relief item included in the Department of Commerce estimates, the reason being that an item of net savings by governmental agencies is subsequently included in our estimate of national income, while this balancing item is absent from the Department of Commerce national income produced totals; (3) our aggregate covers industrial pensions and compensation for injury in but a few industries, whereas the Department of Commerce income paid out covers them fully; (4) among the differences in assumptions underlying the estimates, one deserves singling out, viz., that involved in measuring the entrepreneurial part of income paid out in agriculture. This item in the Department of Commerce estimate is based on the allowance for labor of farm operators and of family members at rates paid to farm workers. In our estimate this amount is raised 25 per cent to allow for the difference in average expenditures on living between farmer and tenant families and farm workers, a ratio established on the basis of scattered sample studies of living expenses on farms.

The quantitative effect of these differences, as well as the other discrepancies between the two sets of estimates, are set forth in Appendix Table IV. Aggregate income payments to individuals exceed income paid out in all years by an amount ranging from 0.7 to 3.0 billion dollars, or from 1 to 7 per cent of the totals. But when the excesses and deficiencies due to the various sources mentioned

[1] See the latest revision in *National Income, 1929–1936*, U. S. Department of Commerce (Washington, 1937).

APPENDIX

Appendix Table IV

COMPARISON OF AGGREGATE INCOME PAYMENTS (N.B.E.R.)
WITH INCOME PAID OUT (D. OF C.), 1929-1935

(absolute figures in millions of dollars)

	1929	1930	1931	1932	1933	1934	1935
Aggregate income payments	79,808	73,620	62,565	49,785	47,880	52,385	56,287
Income paid out	78,174	72,872	61,551	48,487	44,907	51,004	54,645
Gross difference	+1,634	+748	+1,014	+1,298	+2,973	+1,381	+1,642
Accountable excess (+) or deficiency (−) of N.B.E.R. estimate							
Imputed rents	+1,547	+1,378	+953	+614	+557	+459	+521
Relief payments of government (excl. work-relief)					+482	+657	+834
Difference in pensions, compensation for injury, etc.	−127	−134	−123	−122	−144	−179	−240
Difference in entrepreneurial withdrawals in agriculture	+1,130	+1,026	+817	+634	+601	+671	+735
Total of four preceding items	+2,550	+2,270	+1,647	+1,126	+1,496	+1,608	+1,850
Aggregate income payments, adjusted for items above	77,258	71,350	60,918	48,659	46,384	50,777	54,437
Net difference (compared with income paid out)	−916	−1,522	−633	+172	+1,477	−227	−208
Net difference, percentage of income paid out	−1.2	−2.1	−1.0	+0.4	+3.3	−0.4	−0.4

above are taken into account, the residual difference between the two estimates becomes much smaller, ranging from 0.2 to 1.5 billion, and thus at the maximum not exceeding 3.3 per cent of either total. This residual difference appears to be due exclusively to differences in data and procedure; and so far as comparison of the industrial branches is possible, its main locus is in the fields of service, finance and miscellaneous—all fields characterized by paucity of data and hence by the tentative character of the resulting estimates.

Our net savings item is designated as net savings of enterprises; that in the Department of Commerce report is defined as business savings. This difference in designation indicates the difference in scope: our total includes net savings of governmental agencies and the Department of Commerce estimate does not. The other sources of difference lie in the extent to which the adjustment of the accounting measure of net business savings has been carried. In the Department of Commerce report it is confined to the exclusion of business gains and losses on the sale of capital assets. Our adjustment, as mentioned several times, excludes also some of the gains and losses arising from inventory holdings and from the practice of charging capital consumption at original cost rather than at current market prices.

Appendix Table V

COMPARISON OF NET SAVINGS OF ENTERPRISES (N.B.E.R.)
WITH BUSINESS SAVINGS (D. OF C.), 1929-1935

(absolute figures in millions of dollars)

	1929	1930	1931	1932	1933	1934	1935
Net savings of enterprises	3,616	−680	−6,556	−10,157	−8,596	−4,536	−3,252
Business savings	2,583	−4,903	−8,052	−8,942	−3,094	−1,429	310
Gross difference	+1,033	+4,223	+1,496	−1,215	−5,502	−3,107	−3,562
Accountable excess (+) or deficiency (−) of N.B.E.R. estimate							
Savings of government	+1,507	+975	−700	−1,406	−818	−1,181	−1,645
Adjustment for changes in inventory valuation	+712	+4,331	+3,308	+1,520	−2,440	−2,130	−785
Adjustment of depreciation and depletion deductions	−687	−357	+10	+444	+459	+15	−75
Difference in net savings in agriculture	−1,301	−1,325	−1,193	−1,048	−1,053	−1,010	−1,230
Total of four preceding items	+231	+3,624	+1,425	−490	−3,852	−4,306	−3,735
Net savings of enterprises adjusted for items above	3,385	−4,304	−7,981	−9,667	−4,744	−230	483
Net difference (compared with business savings)	+802	+599	+71	−725	−1,650	+1,199	+173

Net savings of enterprises as presented here are compared in Appendix Table V with business savings as estimated in the Department of Commerce report. The gross difference between the two estimates is quite substantial, and varies materially from year to year. But when the effect of the easily ascertainable differences in scope and method is removed, the residual difference is appreciably reduced, becoming negligible in 1931 and rising to a significant figure only in 1933 and 1934.

Appendix Table VI provides a comparison of the most inclusive totals in the two estimates, viz., our national income total and the Department of Commerce national income produced. The gross difference between the two totals ranges from 5.0 billion dollars in 1930, when it is greatest, to a fraction of a billion in 1932, when it is smallest. But the impression that this gross difference creates is misleading. If differences in scope, concept and some of the methods are taken into account, the residual difference becomes appreciably smaller, ranging from 0.04 to 1.0 billion dollars, or from 0.1 to 2.0 per cent of income produced.

Appendix Table VI

COMPARISON OF TOTAL NATIONAL INCOME (N.B.E.R.)
WITH NATIONAL INCOME PRODUCED (D. OF C.), 1929-1935

(absolute figures in millions of dollars)

	1929	1930	1931	1932	1933	1934	1935
Total national income	83,424	72,940	56,010	39,628	39,283	47,849	53,035
National income produced	80,757	67,969	53,499	39,545	41,813	49,575	54,955
Gross difference	+2,667	+4,971	+2,511	+83	-2,530	-1,726	-1,920
Difference accounted for in Table IV	+2,550	+2,270	+1,647	+1,126	+1,496	+1,608	+1,850
Difference accounted for in Table V	+231	+3,624	+1,425	-490	-3,852	-4,306	-3,735
Total of two preceding items	+2,781	+5,894	+3,072	+636	-2,356	-2,698	-1,885
Total national income adjusted for items listed in Tables IV and V	80,643	67,046	52,938	38,992	41,639	50,547	54,920
Net difference (compared with income produced)	-114	-923	-561	-553	-174	+972	-35
Net difference, percentage of income produced	-0.1	-1.4	-1.0	-1.4	-0.4	+2.0	-0.1

APPENDIX C
CHANGES IN PROCEDURE AND SCOPE SINCE THE PUBLICATION OF PRELIMINARY ESTIMATES OF CAPITAL FORMATION IN BULLETIN 52

PRELIMINARY estimates of commodity flow and capital formation were published in Gross Capital Formation, 1919–1933 (*Bulletin 52*, National Bureau of Economic Research, November 15, 1934). The main changes in procedure and scope that have resulted from the succeeding two and a half years of study may be summarized with reference to the measurement of: (1) the flow of finished commodities; (2) the volume of construction; (3) repairs, maintenance and servicing; (4) net changes in commodity inventories.

1 FLOW OF FINISHED COMMODITIES

THREE important stages of the statistical procedure by which we measure the flow of finished commodities to their ultimate domestic recipients are: (a) the ascertainment of the value, at producers' prices, of finished commodities destined for use by domestic ultimate recipients; (b) the estimate of transportation and distributive charges that are to be added to producers' prices in order to measure the flow of commodities at cost to their ultimate users; (c) the measurement of changes in inventories of finished commodities, which allows us to pass from production, adjusted for imports and exports, to sales to ultimate purchasers.

In revising the preliminary estimates, the statistical procedures involved in these three major stages were basically overhauled. In going over the estimates of producers' value of finished commodities, the main distinction between finished and unfinished commodities was checked in the light of published materials and by correspondence with trade associations and commodity experts. The classification of finished commodities by durability was standardized; the interpolation for intercensal years and the adjustment for price changes checked; and the adjustment for imports and exports made more variable over time, in cases of commodities for which other information indicated a marked change in the share of foreign trade.

In the measurement of transportation and distributive charges the changes in procedure were more significant. Transportation costs were, in general, somewhat neglected in preparing the preliminary estimates. More careful consideration led to a marked increase in this item and to a corresponding rise in the final cost to ultimate purchasers. For distributive costs, the former assumption of constancy over time of relative distributive charges (i.e., of gross margins in percentages of the value of sales) was abandoned. On the basis of scattered sample data, an approximate estimate of temporal changes in distributive margins was prepared. And in applying this index of changing distributive margins in each year of the period due attention was paid to the shifting importance of various minor commodity groups within the four major divisions of finished commodities.

The estimate of changes in inventories of finished commodities is part of the general procedure of measuring changes in all commodity inventories. As will be indicated below, the methods of estimating inventory changes were materially revised; in consequence, similar modifications were made in estimating net changes in inventories of finished commodities.

The combined result of all these revisions in the measurement of the flow of finished commodities to ultimate recipients, at cost to them, has been to raise the estimates somewhat, owing primarily to the more inclusive consideration of transportation charges. That this increase does not appear in a direct comparison of the flow of consumers' and producers' durable commodities (Table 10 of this report and Table 5 of *Bulletin 52*) is due to a difference in the scope of the two sets of estimates. The measures in *Bulletin 52* include such repairs and servicing of durable commodities as could be measured with the available data. The estimates in Table 10 of this report exclude them. If this item (Appendix Table VII) is added to the estimates in Table 10, the flow of finished commodities as measured in this report is larger, except in one or two years (exceptions due to differences in estimates of inventory changes and perhaps in classification), than that in *Bulletin 52*.

2 VOLUME OF CONSTRUCTION

In the preliminary estimates the measure of the total volume of construction was based largely upon the estimated consumption of construction materials, raised by a constant ratio of value of construction materials consumed to total value of construction (both expressed in 1929 prices). This particular estimate also has been revised through a more careful and inclusive consideration of the output of construction materials, transportation costs, changes in construction materials inventories, in the cost of distribution of construction materials to their consumers, and in the magnitude of the ratio of construction materials consumed to the total volume of construction. These revisions resulted in a significant increase in the estimated volume of total construction, as may be seen from comparing Appendix Table VII below with Table 5 of *Bulletin 52*.

But in accordance with the interpretation of repair and maintenance construction as an activity whose results are largely non-durable, the volume of construction estimated on the basis of consumption of all construction materials is obviously too inclusive for our purposes. For this reason, another estimate of the volume of new construction, which includes only such substantial repairs and alterations as call for building permits, has been prepared. This estimate was arrived at largely by utilizing the results of other investigators in the field. And it is this estimate of new construction, substantially lower than that of all construction as measured either in *Bulletin 52* or as revised subsequently, that enters the capital formation totals, Variant I or II, in this report.

This change in the area covered by the estimate of construction resulted in an appreciable lowering of the volume of commodity flow and of capital formation. It made possible a more complete classification of gross capital formation by type of user,

Appendix Table VII

VALUE OF REPAIRS, SERVICING AND MAINTENANCE OF DURABLE COMMODITIES, 1919-1933

(millions of dollars)

	1919	1920	1921	1922	1923	1924	1925	1926	1927	1928	1929	1930	1931	1932	1933
								Current Prices							
1 Servicing and repairs of movable durable commodities	2,218	2,641	2,051	2,010	2,456	2,257	2,306	2,423	2,392	2,417	2,576	2,191	1,731	1,367	1,346
a Consumers' durable	361	486	470	455	528	595	660	704	759	832	922	859	788	719	702
b Producers' durable	1,857	2,155	1,581	1,555	1,928	1,662	1,646	1,719	1,633	1,585	1,654	1,332	943	648	644
2 Volume of all construction based on consumption of construction materials	12,158	13,341	11,259	12,626	13,806	14,472	15,664	16,434	16,862	17,385	16,207	14,601	10,200	5,854	5,510
3 Volume of new construction	5,915	6,336	6,105	8,383	9,643	10,491	11,811	11,593	11,787	11,572	10,518	8,629	6,109	3,496	3,230
4 Estimated maintenance and repairs, line 2 - line 3	6,243	7,005	5,154	4,243	4,163	3,981	3,853	4,841	5,075	5,813	5,689	5,972	4,091	2,358	2,280
5 Total repairs, servicing and maintenance, line 1 + line 4	8,461	9,646	7,205	6,253	6,619	6,238	6,159	7,264	7,467	8,230	8,265	8,163	5,822	3,725	3,626
								1929 Prices							
1 Servicing and repairs of movable durable commodities	1,777	1,913	1,923	2,185	2,367	2,303	2,443	2,596	2,478	2,536	2,576	2,268	1,925	1,596	1,584
a Consumers' durable	281	335	364	433	539	614	695	781	822	863	922	912	884	809	799
b Producers' durable	1,496	1,578	1,559	1,752	1,828	1,689	1,748	1,815	1,656	1,673	1,654	1,356	1,041	787	785
2 Volume of all construction based on consumption of construction materials	12,085	10,286	11,384	13,235	13,025	13,822	15,372	16,143	16,629	17,316	16,207	15,006	11,499	7,318	6,711
3 Volume of new construction	5,879	4,886	6,170	8,790	9,100	10,019	11,592	11,388	11,623	11,530	10,518	8,869	6,886	4,372	3,935
4 Estimated maintenance and repairs, line 2 - line 3	6,206	5,400	5,214	4,445	3,925	3,803	3,780	4,755	5,006	5,786	5,689	6,137	4,613	2,946	2,776
5 Total repairs, servicing and maintenance, line 1 + line 4	7,983	7,313	7,137	6,630	6,292	6,106	6,223	7,351	7,484	8,322	8,265	8,405	6,538	4,542	4,360

APPENDIX

and hence a more complete measurement of net capital formation.

3 REPAIRS, MAINTENANCE AND SERVICING

As already indicated, the preliminary estimates include in commodity flow and in capital formation such repairs and servicing of existing durable commodities as can be measured with the data available in the Census of Manufactures and the Census of Retail Trade; and all repairs and maintenance construction covered by the global estimate of the volume of construction based upon the consumption of construction materials. These items are excluded from capital formation as measured in this report.[1] But for possible use by other students in the field, the available estimates of these items are assembled in Appendix Table VII for 1919 through 1933, no attempt having been made to prepare preliminary estimates for 1934 and 1935.

Servicing, repairs and maintenance of durable commodities, to the extent that they could be measured for the period, averaged, in current prices, some 6.9 billion dollars per year, and ranged from a 'low' of 3.6 billion in 1933 to a 'high' of 9.6 billion dollars in 1920. In 1929 prices, the average volume per year amounted to 6.9 billion dollars, ranging from 4.4 billion in 1933 to 8.4 in 1930. The inclusion of this volume would be a significant addition to gross commodity flow; and a very substantial relative addition to gross capital formation, whose average volume over the same period amounted, in current prices to 22.1 billion dollars, and in 1929 prices to 21.5 billion.

The measures in Appendix Table VII must be viewed as crude approximations. The estimates of repairs and servicing of durable commodities are admittedly incomplete since they are confined to services rendered by manufacturing and retail establishments. The year-to-year changes in all the estimates, but especially in the item representing repair and maintenance construction, should not be given much weight. This item is derived as the difference between the global estimate of construction based on the consumption of all construction materials and the estimate of new construction based on substantially different data. Hence it is affected by differences between the assumptions on which the two construction measures are based, and their effect upon the faithfulness with which the two measures reflect fluctuations in the volume of the activities they purport to describe. On the other hand, whatever scanty data are available on repairs and maintenance construction suggest that the average volume of this item in Appendix Table VII is tolerably reliable.

4 NET CHANGES IN COMMODITY INVENTORIES

This element in commodity flow and capital formation is the one for which data are least adequate; and statistical ingenuity can at best produce results that, while plausible, may be vitiated by errors much larger relatively than those possibly present in the other estimates in this report.

The scope and method of measuring net changes in commodity inventories have been altered in several ways since 1934. First, the preliminary estimates evaluated net changes in inventories before 1926, and especially before 1924, on the basis of a regression line of inventory changes on the changes in cost of goods established for corporations since 1925. In the revision these were based on the movement of the inventory-sales ratio for a corporate sample compiled from reports in Moody's and other reference volumes of corporate income accounts and balance sheets. Second, since inventories are estimated on the basis of their relation to the volume of commodity flow, any changes in the estimate of commodity flow would also be reflected in the measure of inventory changes. Third, a more careful and inclusive consideration of the adjustment of current inventories for changes in their valuation affected somewhat the final estimates of net changes. Fourth, we included changes in stocks of monetary metals, an item omitted from the preliminary estimates.

As a result of these modifications, the net change in commodity inventories as estimated in this report differs substantially from that in *Bulletin* 52 (compare Table 10 of this report with Table 5 of *Bulletin* 52). But the whole item is not large as compared with the total commodity flow or capital formation; and since with respect to the direction of change from year to year the present and the earlier estimates are similar, the effect of the revision on the important totals is relatively slight.

This brief account indicates only the major changes in scope and procedure since the publication of the preliminary estimates in 1934. Their combined effect on the total estimate of commodity flow and capital formation was to lower the volumes somewhat and to accentuate their fluctua-

[1] See Section VI for a more detailed discussion.

tions, especially the decline that appeared after 1929. The procedures used and the supporting data will be described in detail in Volume I of *Commodity Flow and Capital Formation*.

APPENDIX D
COMPARABILITY OF ESTIMATES OF CAPITAL FORMATION WITH THOSE OF NATIONAL PRODUCT

NATIONAL income was defined as the net value of commodities and services produced during the year, 'net' in the sense that the total output of all goods is reduced by the value of commodities consumed in the process of production. In order to measure properly this net value, i.e., the net product that can be imputed to the services of individuals and of capital participating in the process of production, the total value produced must be adjusted for the *current* value of the commodities consumed in production. However, it is the practice of business firms, as revealed by standard accounting procedure, to measure costs of production, when they result from the consumption of commodities, not at the market value of the commodities when consumed, but on a different basis, usually at their original cost to the consuming business enterprise. Thus, depreciation and depletion of fixed capital goods are usually reckoned on the basis of original cost to the business enterprise, rather than on the basis of the current reproduction price. For inventories, the principle of cost or market, whichever lower, is followed. As a result, in periods of rising prices inventory consumed is evaluated at cost prices, which are lower than the market price prevailing at the time of consumption; and in periods of declining prices, even though current market prices are used, there is an offsetting loss on the inventory in the profit and loss account. For these reasons we introduced two adjustments to yield an approximation to the net value of commodities and services produced, on the assumption that the cost of currently consumed fixed assets and inventories is evaluated at the current market price, rather than at the cost of these commodities at the time of their purchase by the consuming business enterprise.

But these adjustments, justified as they are in an attempt to arrive at a measure of national product consonant with the theoretical concept, may disturb the comparisons of national income, and hence of gross national product, with our measures of commodity flow and of capital formation. It is therefore important to consider the comparability of the two sets of measures, with reference, first, to the implied evaluation of the current consumption of fixed capital assets; second, to the measurement of the cost of inventories consumed.

Gross capital formation is 'gross' in that the total has not been reduced by the value of fixed capital goods consumed in the process of producing all finished commodities (including unfinished that went into inventories or abroad) or of the goods that enter gross capital formation. But how should this value be calculated, on a book or reproduction price basis? The method does not affect the result of comparisons we wish to make, provided it is consistently applied in deriving the gross national product. If it is assumed that the cost of fixed capital assets consumed should be evaluated on a book value basis, gross national product should be computed by adding to national income, based on the acceptance of the book value basis of capital consumption charges, the total of the latter on a book value basis. If it is assumed that the consumption of fixed capital assets, unadjusted for in gross capital formation, should be calculated on the current reproduction price basis, then the comparable gross national product would be obtained by adding to national income, *adjusted* for the disparity between book and reproduction value bases of capital consumption charges, the capital consumption item estimated on the basis of current reproduction prices. The latter procedure was followed in Table 12 but either treatment would yield the same absolute value of gross national product.

In comparing net capital formation with national income capital consumption charges should likewise be treated consistently. If we obtain net capital formation by subtracting from gross capital formation the volume of capital consumption that is based on the book value of the assets consumed, the comparison should obviously be made

APPENDIX

with national income measured on the same basis (i.e., unadjusted). If total net capital formation is obtained by subtracting from gross capital formation the value of capital consumed at its current market price, the comparison should be made with national income estimated on the same basis, i.e., adjusted for the disparity between the book and reproduction price bases of depreciation and depletion deductions. The second seems to us a more logical approach to the measurement of net capital formation and national income, and has accordingly been adopted in Table 15.

The questions arising in the treatment of the cost of inventories consumed are somewhat more complex. They can best be answered in the form of three brief statements, illustrated by examples.

(1) If it were possible to eliminate completely all gains and losses on holding of inventories, by evaluating inventories consumed at their price current at the time of eventual sale in the form of finished product, then the total national product obtained by adding income payments and savings of enterprises would not be identical with the total obtained by adding consumers' outlay and capital formation. This can be demonstrated by the following oversimplified example. Let us assume two enterprises comprising the whole national economy: A, producing semifinished products and selling the 60 units produced to B, at $1.00 per unit. Enterprise B processes these semifinished products, the sum total of wages, salaries, dividends, etc., including the normal rate of profit, amounting to 42 cents per unit. But the price of the semifinished product rises from $1.00 to $1.25 by the time B is ready to sell the finished product; B then sells the finished product, the total output of the national economy, for $1.67 per unit, or a total of $100. Now if we know that the cost of inventories consumed at the time of eventual sale is $1.25 per unit, our calculation of national income by the method used in this report would be: for enterprise A, $60; for enterprise B, $25 = $100 − (60 × $1.25). The total would thus be $85. The current flow of finished products to ultimate consumers, i.e., consumers' outlay, would, however, be $100, and capital formation, 0, thus yielding by this method a total national product of $100.

(2) But because of limitations of data, the adjustment we made was confined to correcting for the disparity between the current value of the change in inventories and the change in the book value of inventories; no full adjustment was possible for the difference between the price at which inventory commodities were purchased and their price at the time of eventual sale in the form of finished product. Hence the totals of national product obtained by adding net income shares originating should be identical with the totals obtained by adding consumers' outlay and capital formation. Thus, in the example above, there being no inventories at the beginning or end of the time unit, the net saving of enterprise B would remain unadjusted. And the national income total, as obtained in our measurement, would be $60 produced by enterprise A, and $40 = $100 − (60 × $1.00), the total of $100 being the same as that obtained by adding consumers' outlay and capital formation.

This can be further illustrated by introducing inventories and complicating the example somewhat. Let us assume that enterprise B starts with a stock of 60 units of semifinished product bought from A at $1.00 per unit; that enterprise A raises the price to $1.25 on its new output and sells it to B for the purpose of replenishing B's inventories at the end of the year; and that as in the former example B charges $1.67 per unit of its finished product, producing altogether 60 units. On these assumptions, the national income total would be as follows: originating in enterprise A, $75; originating in enterprise B, unadjusted, $100 − ($60 + $75 − $75) = $40; unadjusted national income total, $115; adjustment for the disparity between change in commodity inventories and the difference between successive year-end inventories in current valuation, 0 − $15; adjusted national income total, $100. The same total is obtained by adding consumers' outlay, $100, and capital formation, 0.

It is clear that so long as the volume of inventories is constant there can be no difference in the results obtained by the two methods, even though the book value of inventories changes. For by the net-income-addition method, the net income product of enterprise B will be evaluated at the total sales value of the finished goods it produces minus the price *paid* by B to A for unfinished commodities. The final figure would include no allowance for inventory change, since with a constant volume of inventories, any change in book value will be eliminated by our adjustment. The net income product of enterprise A will be evaluated at the sum it received from enterprise B. The sum of these two must obviously equal the value of the finished goods, since the value of the semifinished goods cancels out. By the consumers' outlay-capital formation method the total is obviously also the value of the finished goods.

(3) This identity between national product,

obtained by summating income payments and savings of enterprises, and that obtained by adding consumers' outlay and capital formation, persists when the commodity volume of stocks changes. Thus, let us retain the conditions of the example just discussed, but assume that enterprise B produces and sells 80 units, reducing its inventory by 20 units. National income will equal: originating in A, $75; originating in B, unadjusted, (80 × $1.67) − ($60 + $75 − $50) = $133.60 − $85 = $48.60; total national income, unadjusted, $123.60 = $75 + $48.60; disparity between change in commodity inventories in current prices [price current during the year is $1.25 per unit, see example under (2)] and difference between successive year-end inventories in changing current valuation, (−20 × $1.25) − ($50 − $60) = −$25.00 + $10 = −$15.00; total national income, adjusted, $123.60 − $15.00 = $108.60. If we calculate by adding consumers' outlay and capital formation, the same result is obtained. Consumers' outlay is 80 × $1.67 = $133.60; capital formation is negative, representing a decline in stocks, and amounts to −20 × $1.25 = −$25.00. Total national product, $133.60 − $25.00 = $108.60.

The only difference between this and the situation under (2) is that in computing the net income product of enterprise B not only must the amount paid to A for unfinished goods be deducted but also the value of the change in inventory (with due regard to signs) must be added. This is what is done by our procedure, by which we first deduct the amount paid to A, add the change in the value of inventory, and then correct the result for the difference between the change in the value of inventory and the value of the change in inventory. It is evident that when the net income product of B, computed in this manner, is added to the net income product of A, the sum must equal consumers' outlay and capital formation, provided the change in inventories is evaluated at the same price in computing both B's net income product and the volume of capital formation.

This brief discussion shows that, theoretically, the measures of the national product, as adjusted by us, should yield results identical with the totals obtained by adding capital formation and consumers' outlay. But this identity can materialize only if statistical difficulties do not prevent the computation of precise measures of the national product totals by the two methods. In actual practice, in the studies that yielded the measures of national income, of gross capital formation, and of capital consumption, a number of assumptions and approximations were made in order to bridge over gaps in the available data. These assumptions and approximations were necessarily different in the different studies and affected the resulting magnitudes, with corresponding effects on the statistical comparability of the measures. How these differences in procedure affect the comparison of the annual estimates may be seen from Appendix Table VIII.

Erratic fluctuations are observable primarily in the *differences* between commodity flow and capital formation, on the one hand, and gross and net national product, on the other; especially when the absolute differences are relatively small, as they are when they represent services not embodied in new commodities. The general effect of these erratic fluctuations, especially in line 11, is to raise the differences in years of depression like 1921, 1924 and 1930. The steps in the statistical procedures used to arrive at the estimates compared in Appendix Table VIII, which explain these erratic fluctuations in the differences, will be stated below in terms of the comparison between commodity flow, including producers' durable commodities, and gross national product. But they are, of course, also applicable to the other comparisons, even though their relative effect on the differences revealed is naturally smaller.

First, the apportionment in Appendix Table VIII is between *cost of services entering* new commodities and those not embodied in new commodities; not between the *quantities* of services, even if weighted by their prices. Hence in a number of industries supplying jointly both producers of new commodities and ultimate consumers or producers of services not embodied in new commodities, the cost attributable to the production and distribution of new commodities is likely to rise and decline with business cycles, much more than would the volume of services at current prices. Thus, in the case of government, insurance and banking—all branches whose net product is quite unresponsive to business cycles—the share of their gross and net product attributed to and entering as cost to the producers, transporters and distributors of commodities fluctuates, of course, with business cycles. (This share would be represented by business taxes, short term interest payments, some dividend and long term interest payments made by the enterprises to banks and insurance companies.) Similarly, the contributions by producers, transporters and distributors of new commodities out of the cost of these commodities to the maintenance of semipublic enterprises (hos-

Appendix Table VIII

ANNUAL ESTIMATES OF NATIONAL PRODUCT, CAPITAL FORMATION AND CONSUMERS' OUTLAY, 1919-1935

(millions of dollars)

	1919	1920	1921	1922	1923	1924	1925	1926	1927	1928	1929	1930	1931	1932	1933	1934	1935
1 Gross national product	68,750	82,836	66,148	67,186	78,214	78,791	83,413	88,780	86,778	90,053	93,640	82,723	64,751	47,202	46,538	55,765	61,243
2 Gross capital formation, Variant I	19,341	22,100	11,488	13,282	18,199	15,245	19,211	19,037	18,208	17,824	20,298	13,662	8,464	3,147	4,268	6,061	9,008
3 National income	59,926	72,386	58,343	59,706	69,706	70,369	74,846	79,477	77,429	80,397	83,424	72,940	56,010	39,628	39,283	47,849	53,035
4 Net capital formation, inclusive total	10,517	11,650	3,683	5,802	9,691	6,823	10,644	9,734	8,859	8,168	10,082	3,879	−278	−4,427	−2,987	−1,855	800
5 Consumers' outlay, line 1 − line 2 or line 3 − line 4	49,409	60,736	54,660	53,904	60,015	63,546	64,202	69,743	68,570	72,229	73,342	69,061	56,288	44,055	42,270	49,704	52,235
6 Consumers' outlay	49,409	60,736	54,660	53,904	60,015	63,546	64,202	69,743	68,570	72,229	73,342	69,061	56,288	44,055	42,270	49,704	52,235
7 Perishable commodities	24,646	27,278	22,047	21,410	22,967	23,750	25,404	27,107	26,672	27,348	28,550	26,395	21,481	18,147	18,133	20,756	23,095
8 Semidurable commodities	10,451	12,156	9,736	10,023	11,324	10,735	11,361	11,917	12,032	12,193	12,382	10,731	9,024	6,722	6,513	7,512	8,151
9 Consumers' durable commodities	5,987	6,921	5,570	6,181	7,943	7,900	9,056	9,445	8,890	9,174	9,913	7,550	5,748	3,806	3,882	4,686	5,918
10 Total commodities, line 7 + line 8 + line 9	41,084	46,355	37,353	37,614	42,234	42,385	45,821	48,469	47,594	48,715	50,845	44,676	36,253	28,675	28,528	32,954	37,164
11 Services not embodied in new commodities, line 6 − line 10	8,325	14,381	17,307	16,290	17,781	21,161	18,381	21,274	20,976	23,514	22,497	24,385	20,035	15,380	13,742	16,750	15,071

pitals, etc.) are likely to fluctuate with business cycles. As a result a large share of the product of these service industries is embodied, in years of business prosperity, as cost in new commodities and a smaller share imputed to services not so embodied, while in years of business depression the reverse is true.

Second, the factor just mentioned, which in itself would go far to explain the rise in the value of services not embodied in new commodities in years of depression, is magnified by certain peculiarities of the estimate of national income, and hence of gross national product. The scantiness of data makes possible only rough approximations to the net value produced in several industrial divisions; and the crudity of these approximations means in general that the resulting estimates do not reflect sensitively the fluctuations that may occur. It so happens that this lack of data and the resulting insensitiveness of the estimates is particularly predominant in finance, service and miscellaneous, i.e., exactly those fields in which services are rendered jointly to producers of new commodities and to ultimate consumers and producers of services not embodied in new commodities. Since the annual estimates of capital formation are less subject to this weakness, it is quite possible that they reflect cyclical fluctuations in the areas they are supposed to measure much more sensitively than our estimates of national income and of gross national product reflect cyclical fluctuations in the final product of the economic system.

Third, in the commodity classification underlying the measurement of capital formation, several commodities were classified as finished because only minor fractions of them were consumed by business enterprises. These fractions may be consumed by enterprises producing other commodities. Such duplication is offset somewhat by the classification as unfinished of some commodities that may, to a very small extent, be consumed directly by ultimate consumers. But the important point is that the extent of duplication or deficiency in the finished commodity totals is subject to a definite cyclical change. This is a result of the fact that consumption of finished commodities by enterprises producing other finished commodities is much more sensitive to business cycles than is consumption of unfinished commodities by ultimate consumers.

Fourth, in the estimate of capital formation we assumed that inventories held by manufacturing establishments are predominantly unfinished commodities, and we did not allow for changes in them in estimating the flow of finished commodities to ultimate consumers. So far as manufacturers' inventories do include finished commodities and so far as these inventories tend to rise and decline with business cycles, the effect would be to overestimate the flow of finished commodities during years of business prosperity and to underestimate it in years of business depression.

The characteristics indicated above suggest why the differences representing the value of services not embodied in new commodities tend to be lower than would be expected in years of business expansion, and to rise in years of business contraction. This tendency may, however, be offset somewhat by one characteristic still to be mentioned. The estimates of national income and of gross national product should exclude gains and losses of business enterprises on the sale of capital assets; and they do exclude these items since 1929. But for the earlier years the available data do not allow this adjustment.

These various deficiencies of the estimates compared in Tables 12, 15 and 16 provide sufficient reason for avoiding comparisons in terms of unsmoothed annual data and for disregarding the year-to-year changes that such comparisons would reveal. But it should be noted in conclusion that while these deficiencies are likely to disturb significantly the movement of such small differences between two large totals as appear in line 11 of Appendix Table VIII and in line 4 of Table 16 they are not likely to affect seriously such large totals as those of gross or net national product, commodity flow or gross capital formation; and are not likely to disturb significantly the average magnitude of net capital formation or the striking changes over time that are observed in its volume.

PUBLICATIONS OF THE
NATIONAL BUREAU OF ECONOMIC RESEARCH, INC.

PUBLICATIONS OF THE NATIONAL BUREAU OF ECONOMIC RESEARCH

*1 INCOME IN THE UNITED STATES
W. C. Mitchell, W. I. King, F. R. Macaulay and O. W. Knauth;
Volume I (1921) Summary 152 pp.
2 Volume II (1922) Details 440 pp., $5.15
3 DISTRIBUTION OF INCOME BY STATES IN 1919 (1922) 30 pp., $1.30
Oswald W. Knauth
*4 BUSINESS CYCLES AND UNEMPLOYMENT (1923) 405 pp.
By the National Bureau Staff and sixteen Collaborators
*5 EMPLOYMENT, HOURS AND EARNINGS, UNITED STATES, 1920–22 (1923) 147 pp.
Willford I. King
6 THE GROWTH OF AMERICAN TRADE UNIONS, 1880–1923 (1924) 170 pp., $2.50
Leo Wolman
7 INCOME IN THE VARIOUS STATES: ITS SOURCES AND DISTRIBUTION, 1919, 1920 AND 1921 (1925)
Maurice Leven 306 pp., $3.50
8 BUSINESS ANNALS (1926) 380 pp., $2.50
By Willard L. Thorp, with an introductory chapter, Business Cycles as Revealed by Business Annals, by Wesley C. Mitchell
9 MIGRATION AND BUSINESS CYCLES (1926)
Harry Jerome 256 pp., $2.50
10 BUSINESS CYCLES: THE PROBLEM AND ITS SETTING (1927) 489 pp., $5.00
Wesley C. Mitchell
*11 THE BEHAVIOR OF PRICES (1927) 598 pp.
Frederick C. Mills
12 TRENDS IN PHILANTHROPY (1928) 78 pp., $1.00
Willford I. King
13 RECENT ECONOMIC CHANGES (1929) 2 vol., 950 pp., $7.50
By the National Bureau Staff and fifteen Collaborators
14 INTERNATIONAL MIGRATIONS
Volume I, Statistics (1929), compiled by Imre Ferenczi of the International Labour Office, and edited by Walter F. Willcox 1,112 pp., $7.00
18 Volume II, Interpretations (1931), edited by Walter F. Willcox 715 pp., $5.00
*15 THE NATIONAL INCOME AND ITS PURCHASING POWER (1930) 394 pp.
Willford I. King
16 CORPORATION CONTRIBUTIONS TO ORGANIZED COMMUNITY WELFARE SERVICES (1930)
Pierce Williams and Frederick E. Croxton 347 pp., $2.00
17 PLANNING AND CONTROL OF PUBLIC WORKS (1930) 260 pp., $2.50
Leo Wolman
*19 THE SMOOTHING OF TIME SERIES (1931) 172 pp.
Frederick R. Macaulay

20 THE PURCHASE OF MEDICAL CARE THROUGH FIXED PERIODIC PAYMENT (1932)
Pierce Williams 308 pp., $3.00
*21 ECONOMIC TENDENCIES IN THE UNITED STATES: ASPECTS OF PRE-WAR AND POST-WAR CHANGES (1932) 639 pp.
Frederick C. Mills
22 SEASONAL VARIATIONS IN INDUSTRY AND TRADE (1933) Simon Kuznets 455 pp., $4.00
23 PRODUCTION TRENDS IN THE UNITED STATES SINCE 1870 (1934) Arthur F. Burns 363 pp., $3.50
24 STRATEGIC FACTORS IN BUSINESS CYCLES (1934)
John Maurice Clark 238 pp., $1.50
25 GERMAN BUSINESS CYCLES, 1924–1933 (1934)
Carl T. Schmidt 288 pp., $2.50
26 INDUSTRIAL PROFITS IN THE UNITED STATES (1934) Ralph C. Epstein 678 pp., $5.00
27 MECHANIZATION IN INDUSTRY (1934)
Harry Jerome 484 pp., $3.50
28 CORPORATE PROFITS AS SHOWN BY AUDIT REPORTS (1935) William A. Paton 151 pp., $1.25
29 PUBLIC WORKS IN PROSPERITY AND DEPRESSION (1935) Arthur D. Gayer 460 pp., $3.00
30 EBB AND FLOW IN TRADE UNIONISM (1936)
Leo Wolman 251 pp., $2.50
31 PRICES IN RECESSION AND RECOVERY (1936)
Frederick C. Mills 561 pp., $4.00
32 NATIONAL INCOME AND CAPITAL FORMATION, 1919–1935 (1937) Simon Kuznets 100 pp., 8¼ x 11¾, $1.50

STUDIES IN FINANCE
A PROGRAM OF FINANCIAL RESEARCH (1937)

I Report of the Exploratory Committee on Financial Research 91 pp., $1.00
II Inventory of Current Research on Financial Problems 253 pp., $1.50

CONFERENCE ON RESEARCH IN NATIONAL INCOME AND WEALTH

STUDIES IN INCOME AND WEALTH, Volume I (1937) 368 pp., $2.50

IN PRESS

COMMODITY FLOW AND CAPITAL FORMATION, Volume I Simon Kuznets 320 pp., 8¼ x 11¾, $5.00
SOME THEORETICAL PROBLEMS SUGGESTED BY THE MOVEMENTS OF INTEREST RATES, BOND YIELDS AND STOCK PRICES IN THE UNITED STATES SINCE 1856 500 pp., $5.00
Frederick R. Macaulay

THE BULLETIN
Subscription to the National Bureau Bulletin (5 issues, $1). Single copies, 25¢, except for the issues marked 50¢.

1934
49 NATIONAL INCOME, 1929–1932, Simon Kuznets
50 RECENT CORPORATE PROFITS, Solomon Fabricant
51 RECENT CHANGES IN PRODUCTION, Charles A. Bliss
52 GROSS CAPITAL FORMATION, 1929–1933, Simon Kuznets 50¢
53 CHANGES IN PRICES, MANUFACTURING COSTS AND INDUSTRIAL PRODUCTIVITY, 1929–1934, Frederick C. Mills

1935
54 WAGES AND HOURS UNDER THE CODES OF FAIR COMPETITION, Leo Wolman
55 PROFITS, LOSSES AND BUSINESS ASSETS, 1929–34, Solomon Fabricant
56 ASPECTS OF MANUFACTURING OPERATIONS DURING RECOVERY, Frederick C. Mills 50¢
57 THE NATIONAL BUREAU'S MEASURES OF CYCLICAL BEHAVIOR, Wesley C. Mitchell and Arthur F. Burns 50¢
*58 PRODUCTION IN DEPRESSION AND RECOVERY, Charles A. Bliss

1936
59 INCOME ORIGINATING IN NINE BASIC INDUSTRIES, 1919–1934, SIMON KUZNETS 50¢
60 MEASURES OF CAPITAL CONSUMPTION, 1919–1933, SOLOMON FABRICANT
61 PRODUCTION DURING THE AMERICAN BUSINESS CYCLE OF 1927–1933, WESLEY C. MITCHELL and ARTHUR F. BURNS 50¢
62 REVALUATIONS OF FIXED ASSETS, 1925–1934, SOLOMON FABRICANT
63 THE RECOVERY IN WAGES AND EMPLOYMENT, LEO WOLMAN

1937
64 A PROGRAM OF FINANCIAL RESEARCH
Report of the Exploratory Committee on Financial Research
65 NON-FARM RESIDENTIAL CONSTRUCTION, 1920–1936, DAVID L. WICKENS and RAY R. FOSTER
*66 NATIONAL INCOME, 1919–1935, SIMON KUZNETS
67 TECHNICAL PROGRESS AND AGRICULTURAL DEPRESSION, EUGEN ALTSCHUL and FREDERICK STRAUSS 50¢
68 UNION MEMBERSHIP IN GREAT BRITAIN AND THE UNITED STATES, LEO WOLMAN

* *Out of print.*

NATIONAL BUREAU OF ECONOMIC RESEARCH

1819 BROADWAY, NEW YORK

EUROPEAN AGENT: MACMILLAN & CO., LTD.

ST. MARTIN'S STREET, LONDON, W. C. 2

Soc
HC
106.3
K8